STO

**ACPL ITEM
DISCARDED**

7-6-70

Geophysics

Series Editor R. F. Kempa PhD

Geophysics

Walter Shepherd

With line drawings by the author

G. P. PUTNAM'S SONS
200 MADISON AVENUE NEW YORK

Copyright © 1969 by Walter Shepherd
All rights reserved. No part of this publication may be reproduced, stored in a retrieval system, or transmitted, in any form or by any means, electronic, mechanical, photocopying or otherwise, without the prior permission of the copyright owner.

Library of Congress Catalog Card Number: 72-93753

Acknowledgments

The author and publishers wish to express their thanks to the United States Information Service, and British Aircraft Corporation Ltd, for permission to reproduce copyright photographs in this book.

Photoset by BAS Printers Limited, Wallop, Hampshire, England
Printed in Great Britain by The Stellar Press Ltd., Hatfield, Herts

Contents

	page
1 What is Geophysics?	7
International Cooperation	8
International Geophysical Year	11
Applied Geophysics	13
2 The Planet Earth	16
The Nature of the Earth	18
The Shape of the Earth	21
The Origin of the Earth	25
The Earth's Heat	26
Chemical Composition	29
3 The Lithosphere	33
Mountain Building	33
Floating Continents	36
Continental Drift	39
Has the Earth Expanded?	47
The Interior of the Earth	48
The 'Mohole' Project	53
Vulcanicity	54
4 The Hydrosphere	57
The Sea Floor	58
Conditions under the Sea	62
Ocean Currents	64
The Tides	67
Tidal Currents	72
Deep-sea Sounding	73
Exploration Downward	75
Into the Depths	80

5 The Atmosphere and Beyond 84
   The Troposphere 85
   Ferrel's Law 89
   Further Modifications 92
   Exploration Upward 94
   The Stratosphere 96
   Probing with Waves 102
   The Ionosphere 104
   The Chemosphere 105
   The Exosphere 106
   Meteors and Cosmic Rays 108
   Dynamo Currents 109
   The Magnetosphere 109
   The Aurora 113

Appendix – The Biosphere 117

Index 123

1

What is Geophysics?

Geophysics is the study of the characteristic properties of the earth as a whole. It attempts to answer such questions as these: What is the earth's shape, its size, its motion? What is it made of, what is its structure, and what is the state of its interior? Geophysics is also concerned with the general character of the earth's surface, the physical properties of its atmosphere, and its relations with surrounding space. It thus involves studies in chemistry, geology, meteorology, astronomy, and many other subjects, but its distinguishing feature is that it seeks to describe the earth as a single object rather than as a collection of parts. It is man's attempt to understand his world, and he calls it the 'natural science of the earth' from the Greek *ge*, the earth, and *physike*, natural science.

The astronomers were the first to consider the earth in this way. They produced various theories of its origin, shape and size, and the geologists followed with views based on the evidence of the rocks. The traditional legend of the Flood set them looking for signs of ancient inundations, and the discovery of enormously thick sediments, which must have taken a very long time to settle, raised the question of the earth's age. Volcanoes, again, suggested nether fires and roused interest in the earth's interior, while the laws of gravity led to the question of the weight of the earth.

A great many of these questions were – and still are – investigated by individual researchers, but problems presently arose which could be solved only by knowing what was happening in several parts of the earth at the same time. This often required team work involving observers stationed in different countries, and geophysics today is very largely dependent on international collaboration. Cooperation of this kind was difficult to organize effectively until the telegraph and radio were invented, but the idea is by no means new. 'Experiments in concert' were suggested 350 years ago in England by Francis Bacon, though nothing much of this sort was attempted until the end of the 18th century.

In the 1790's the chemist Antoine Lavoisier, the biologist Jean Baptiste Lamarck and the astronomer Pierre Simon Laplace planned a chain of weather stations through France, and from 1800 to 1815 regular reports (of past weather) were published for scientific study. In the 1850's similar reports, whose transmission was speeded up by the invention of the electric telegraph in 1837, were organized (independently) on a national scale in Great Britain and America, this time with the purpose of providing weather information for the public.

International Cooperation

In 1851 Matthew Maury, an American naval officer, suggested international co-operation in the preparation of weather reports from all over the world for the benefit of navigators. As a result, the first International Meteorological Congress was held in Brussels in 1853. Ten countries were represented, and the possibilities of forecasting the weather were explored. This led, in 1878, to the formation of the International Meteorological Organization, the forerunner of the present World Meteorological Organization at Geneva. This body is mainly concerned with weather forecasting, which we must regard as a branch of applied geophysics too specialized for treatment here.

Meanwhile, the German scientist and explorer, Alexander von Humboldt, had secured the cooperation of Russia, Great Britain, America and a number of other countries in a worldwide magnetic survey. Data were supplied from stations scattered through the Russian and British empires, from America as far north as Alaska, and from China. At the same time, the mathematician Karl Gauss and the physicist Wilhelm Weber founded the Magnetic Union at Göttingen, Germany, and obtained international cooperation in making simultaneous magnetic observations at widely separated places on specified dates.

The work began by spanning Europe from the Netherlands to Sicily from 1835 to 1839, and as its importance was realized many observatories started to keep clocks set to Göttingen time to facilitate recording. In France, the International Association of Geodesy was established in 1864 for the purpose of determining the exact shape of the earth, and at Gratz, in 1875, Karl Weyprecht, an Austrian naval lieutenant and Arctic explorer, read a paper to the German Scientific and Medical Association which led to the First Polar Year.

Weyprecht suggested that chains of stations should be established round the Arctic and Antarctic Circles to record weather data, magnetic phenomena, the aurora, and the behavior of the polar ice. Detailed plans were prepared for submission to the International Meteorological Congress which met in Rome in 1879, and in Germany Bismarck appointed a commission of scientists to report on the scheme. In 1880 an International Polar Conference met in Hamburg, where eight countries were represented. At a second conference, held at Berne, an International Polar Commission was set up to organize a program. This included the establishment of twelve polar stations in the Arctic and two in the Antarctic, the observations to be carried out for one year starting in August, 1882.

The Antarctic stations were on land, one at Cape Horn and the other in South Georgia, but of the more hazardous Arctic expeditions, two met with disaster. The Dutch ship bound for the mouth of the river Yenisei, in Siberia, was crushed by ice, and the American expedition to Ellesmere Island, under Lieutenant Adolphus Greely, ran short of provisions during 1882 and their relief ship failed to reach them. In 1883 they set out in small boats from their camp on the ice, in a heroic effort to reach the land, but they were trapped by floes. They had to abandon all but one boat, and this they were obliged to drag across the ice. Twice, the floe supporting them was driven back into the polar sea by storms, but after fifty-one exhausting days they reached land expecting to find a rescue ship awaiting them. This ship, however, had also been crushed by ice, its crew escaping to Greenland, and all they found was a small cache of stores. They were compelled to face a third Arctic winter, during which seventeen of their twenty-five members died of starvation and exposure.

The devotion to duty of the men of both these expeditions was itself a phenomenon. Those marooned on the ice by the loss of the Dutch ship camped on the floes and maintained their scientific observations through severe hardship until they were finally able to reach land. The American expedition described above, having reached a point nearer the North Pole than had ever been reached before, conscientiously completed its observations under starvation conditions. Finally, abandoning all hope of rescue, the men made duplicate copies of all their records, secured them in waterproof containers, and placed them in separate boats to halve the chance of loss. They continued with their observations, taking barometric and magnetic readings, and noting auroral displays, until they were too weak to do more than note the weather, and they persisted in this long after the Polar Year had ended. The eight survivors were at last rescued by a search party from a ship, but one died on the way home and another

What is Geophysics?

had lost both his hands and feet.

The Second Polar Year was first seriously considered in 1927, and elaborate plans were made at a conference at Christiansborg, Denmark, in 1928. It was suggested that airships might be used to set up stations and keep them supplied, and four stations were proposed in Antarctica. A special commission for the Polar Year 1932–1933 proposed that the plans should be widened to include the earth as a whole. Sounding balloons were to be employed to gather data from the upper atmosphere, and there were to be special recordings of magnetic phenomena and earth currents during the eclipse of the sun on August 31, 1932. Cameras were to be used to photograph the auroras, and experiments were to be made in radio transmission round the curvature of the earth. At a conference in Leipzig in 1929 it had been suggested that ground stations should record the sound of successive explosions high in the atmosphere, and it was now proposed in America that Robert H. Goddard's new liquid-fuel rockets should be used.

This ambitious program had to be curtailed for lack of funds, but the required magnetic and other instruments, including radio sondes (see page 94), were provided with the help of grants from the Rockefeller Foundation in New York. When the Second Polar Year began, 22 countries were sending out expeditions beyond their own borders, and 30 magnetic stations were established in the Arctic and sub-Arctic regions. The publishing and collation of the vast amount of information gathered took several years and was interrupted by World War II, during which the maps containing the data for the last fifteen days of the Second Polar Year were lost.

International Geophysical Year

The information gathered during the Polar Years was immensely valuable, especially in the field of radio communications, and in 1951 the International Council of Scientific Unions, representing some 45 nations, met in Washington to plan the Third Polar Year. This project was not, however, intended to be exclusively 'polar.' At Amsterdam in 1953 it was agreed that it should be called the International Geophysical Year (IGY), and that it should run from July 1, 1957, to December 31, 1958, a period of 18 months.

When the IGY opened, some 60,000 scientists from 66 nations were established in thousands of stations all over the world. There were several in Antarctica and one at the South Pole. The discoveries made included the Van Allen radiation belts surrounding the earth. The technological triumphs included the successful launching of *Sputnik I*, the first artificial satellite, but the enormous success of the IGY cannot be gauged by listing the outstanding achievements, for it includes a wealth of detailed information gathered from oceanographic surveys, weather reports from remote regions, countless magnetic and auroral observations, recordings sent back from the upper atmosphere by radio sondes and rockets, measurements of earthquake shocks, observations of volcanic activity, surveys of the Antarctic and other little-known regions, observations of the sun and moon, the tracking of artificial satellites and so on.

The IGY was a period of maximum sunspot activity, and it has since been complemented by International Years of the Quiet Sun (IQSY), devoted to the study of phenomena during the periods of minimum sunspot activity. In the IQSY which ended in 1965, 71 nations manned 2,054 geophysical stations. International cooperation has now become a common feature of geophysical research, and other periodic activities include the reporting of magnetic phenomena on five International Quiet Days, and five International Disturbed Days, every month. Daily reports on the state of the atmosphere at different levels are also broadcast round the world from ground

What is Geophysics?

stations, weather-ships and artificial satellites. Further, there are several international organizations concerned with space research and telecommunications, and a number of permanent geophysical stations are now maintained in the Antarctic by many nations.

The oceans, too, have become open fields for international research, particularly since the greater part of the sea floor is not national property. The first full-scale conference on submarine exploration was held in 1969 at Brighton, England, under the title 'Oceanology International '69', and 14 nations were represented.

Applied Geophysics

The main practical uses of geophysics may be grouped under the following heads: prospecting for minerals, forecasting the weather, and developing telecommunications. Other applications may be found as the results of future research, including the prediction of earthquakes and volcanic eruptions. Of the three subjects, only the first needs brief mention here, since there is abundant popular literature on the other two.

The methods employed in geophysical prospecting fall into five groups: the magnetic, the electrical, the gravity, the seismic, and the radiometric. The simplest and oldest is the magnetic method, which consists essentially in searching for iron ore with a compass needle. The earth's magnetic field is usually strong enough to smother the effects of small local masses of iron ore on ordinary compasses, unless they are very close, though masses of the magnetic oxide ('lodestone') can be readily detected in this way. Nevertheless, very sensitive magnetometers have been devised which will show the presence of iron ores covered by 5,000 feet or more of other rocks, and successful magnetic surveys have been made by flying over the land in an airplane.

The electrical methods include the detection of the

minute currents generated naturally in the soil surrounding some ore deposits; the detection of the distortions produced by metallic ores in an artificial electric field produced in the ground by means of buried electrodes; and the measurement of the change in the electrical resistance of the ground caused by ore-bodies. Radio waves are also used, either to detect the presence of conducting bodies by their echoes (as in radar), or in the highly successful method known as 'inductive' prospecting, which is done from an airplane.

In this, high frequency electromagnetic pulses are sent out from loop aerials connecting the nose, wing tips and tail of the airplane. The pulses penetrate 500 feet or more into the ground, where they induce eddy currents in any conducting ore bodies. The electromagnetic fields caused by these currents are then detected by sensitive receiving coils in a small bomb-shaped device called a 'bird,' which is slung beneath the airplane on a long wire.

Electrical methods will also reveal underground water, and as long ago as 1914 water-levels in South-West Africa were found by trailing a long wire from an airplane, the wire and the hidden water forming a condenser whose capacity could be measured. Both magnetic and electrical (soil-conductivity) techniques were employed by archaeologists in 1967 in the survey of Cadbury Castle, near Yeovil, Somerset. By these means numerous infilled storage pits, post holes, trenches and collapsed walls dating from the 5th century were located.

Gravity measurements began with the pendulum experiments referred to on page 36, but since then very sensitive torsion balances have been devised which can detect the gravitational attraction of a few tons of coal in a cellar. The method in prospecting is to seek anomalies due to hidden masses of greater or less density than the 'country rock,' but the nature of the discovered masses has to be ascertained by other means. Ore deposits, salt domes, water, and certain types of geological formation

What is Geophysics?

have been detected by this method, sometimes at more than a mile below the surface.

The seismic method is commonly used in prospecting for oil, although it shows only whether the subterranean geological structure is favorable to the accumulation of oil and does not detect the oil itself. A small charge of explosive is fired in the ground, and the vibrations are picked up by seismographs situated at various distances from it. The surface of a buried stratum of different density from the overlying rock reflects the seismic waves back to the surface, and the data gathered enable its depth and form to be ascertained. The method is similar to that described in Chapter 3 for making earthquake waves reveal the structure of the earth's interior. During the International Geophysical Year it was used to ascertain the thickness of the ice sheet over the Antarctic continent. It has also enabled the land surface of Greenland to be mapped, and has shown that, beneath its ice cap, Greenland consists of two large islands, not one.

The radiometric method is used in prospecting for radioactive minerals emitting gamma rays. If the deposits are fairly near the surface the rays can be picked up by scintillation detectors carried in airplanes flying at 500 feet or less. The type of mineral emitting the rays could not be distinguished until a new type of gamma-ray spectrometer was invented in 1968. This enables the three most important radioactive elements – uranium, thorium and potassium – to be identified. The gamma-ray method has also proved useful in detecting deposits of phosphates, and in geological surveying, while the suspected occurrence of radioactive haloes around oil deposits has suggested its possible use in oil prospecting.

2

The Planet Earth

The earth is one of the smaller members of the solar system, and to an outside observer (should there be one!) it is probably one of the least important. But to man it was for thousands of years the center of the universe, and it is really astonishing that he recognized it as but a small member of a large family of planets as early as he did. That it is a spherical ball was taught by the Pythagoreans more than 2,500 years ago, and that the earth, the sun and the moon are all made of the same materials, the sun being 'red hot' and the moon shining by its reflected light, was maintained in 450 BC by Anaxagoras, who also explained eclipses correctly.

Aristotle (*c.* 330 BC) maintained that the earth occupies the center of the universe, which was an error, but this pupil Heraclides correctly explained day and night by the rotation of the earth on its axis, and understood that Venus and Mercury are planets revolving about the sun. That the earth also revolves about the sun, and the moon about the earth, was held in 270 BC by Aristarchus, who gave a satisfactory account of the phases of the moon. At about 240 BC Eratosthenes calculated the circumference of the earth from careful observations and was less than 200 miles in error.

These conclusions were reached by reasoning from naked-eye observations, but they contradicted ancient beliefs and were contested on the grounds of religion, tradition and 'common sense.' Thus, the old mistake that

The Planet Earth

the earth lies in the center of the universe was taught by the great astronomical observers Hipparchus (190—120 BC) and Ptolemy of Alexandria (*c.* AD 150), and this doctrine remained 'official' in Christendom through the Middle Ages. The earth was not to be 'put in its place' again until the 16th century, when Nicolaus Copernicus (1473–1543) gave the first true picture of the solar system for more than 1,800 years. It met with the same sort of

Table I: THE EARTH

DIMENSIONS

Mean radius	3,956·5 miles
Equatorial diameter	7,926·7 miles
Polar diameter	7,901 miles
Mean diameter, sun = 1	1/110
Circumference at equator	24,902·5 miles
Surface area	$1·96 \times 10^8$ sq. miles
Land	29%
Water	71%

DISTANCE FROM SUN

Maximum (21 June)	94,500,000 miles (approx.)
Minimum (21 December)	91,414,000 miles (approx.)
Mean (as measured in 1967)	92,957,000 miles (approx.)

ORBIT

Mean diameter	186,000,000 miles (approx.)
Period of revolution (1 year)	365·25 days
Mean orbital velocity	66,579 m.p.h. = 18·5 m.p.sec.

ROTATION

Period (1 day)	24 hr. 56 min. 4·09 sec.
Velocity at equator	1,040 m.p.h.
Velocity at latitude of London ($51\frac{1}{2}°$)	647 m.p.h.
Inclination of axis to ecliptic	66° 32'

MASS AND GRAVITY

Mass	$5·88 \times 10^{21}$ tons
Mass, sun = 1	1/332,260
Density, water = 1	5·527
Acceleration due to gravity	32 ft./sec.^{2}
Escape velocity	7 m.p.sec.

opposition as before, being rejected by such great observers as Tycho Brahe (1546–1601), but was finally established by the work of Galileo Galilei (1564–1642), Johannes Kepler (1571–1630), and Isaac Newton (1642–1727). The weight of the earth was first determined with near accuracy by Henry Cavendish (1731–1810), the accepted figure being almost exactly 6,000,000,000,000,000,000,000 tons.

The invention of the telescope by Johannes Lippershey in 1608 led eventually to the construction of astronomical and surveying instruments of great precision, and by the end of the 19th century the form, dimensions and motions of the earth were known to a high degree of accuracy. Later refinements have been proportionately very minute, and the statistics given in Table I are regarded as thoroughly well established and precise enough for all practical – and most theoretical – purposes.

The Nature of the Earth

The earth is often considered as a ball of rock surrounded by a series of concentric shells, and this provides a convenient division of its study into three major parts. The solid ball of the earth is called the *lithosphere*, and nearly three-quarters of its surface is covered by the water of the *hydrosphere*. The hydrosphere is extremely thin compared with the lithosphere. If the circle of the earth were drawn on the edges of a pile of 3,400 sheets of paper (about 7 inches thick), the top and bottom sheets would represent the average depth of the hydrosphere. The great ocean deeps would penetrate through three sheets of paper in a few isolated places.

Surrounding the lithosphere and the hydrosphere is the envelope of mixed gases called the *atmosphere*. This is of indefinite thickness since it has no outer boundary, but it becomes negligible at an altitude of about 500 miles. It is difficult to conceive the rarity of the gases at such a

The Planet Earth

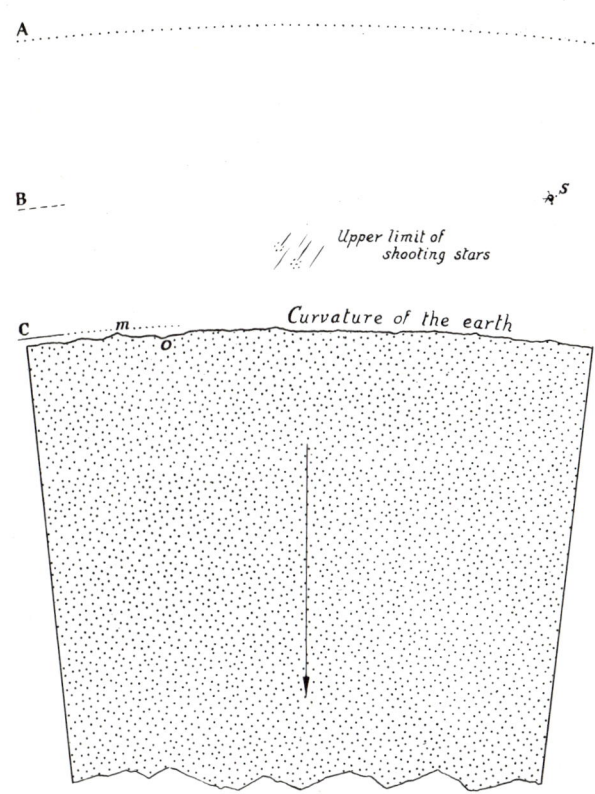

Fig. 1. Part of a sector of the earth: scale approximately 1: 10,000,000; radius of curvature 13 inches, center in direction of arrow. A, Atmosphere an ultra-vacuum. B, Lower limit for all but very short-lived artificial satellites (s). C, Upper limit of clouds and weather; bottom of stratosphere; air too thin to breathe. *m* Highest mountains. *o* Deepest oceans.

height, for the atmosphere here (represented by about 200 sheets of paper) would be called an ultra-high vacuum at the surface of the earth, and a vessel containing it would normally be regarded as completely empty. The depth of breathable air would be about the thickness of only one sheet of paper.

An idea of the comparative thicknesses of the three principal shells of the earth is given in the section in Fig. 1, but there is also a somewhat different type of zone known as the *biosphere*. This term is used to denote the narrow shell within which life normally exists. It ranges locally from the ocean deeps to about the highest mountain

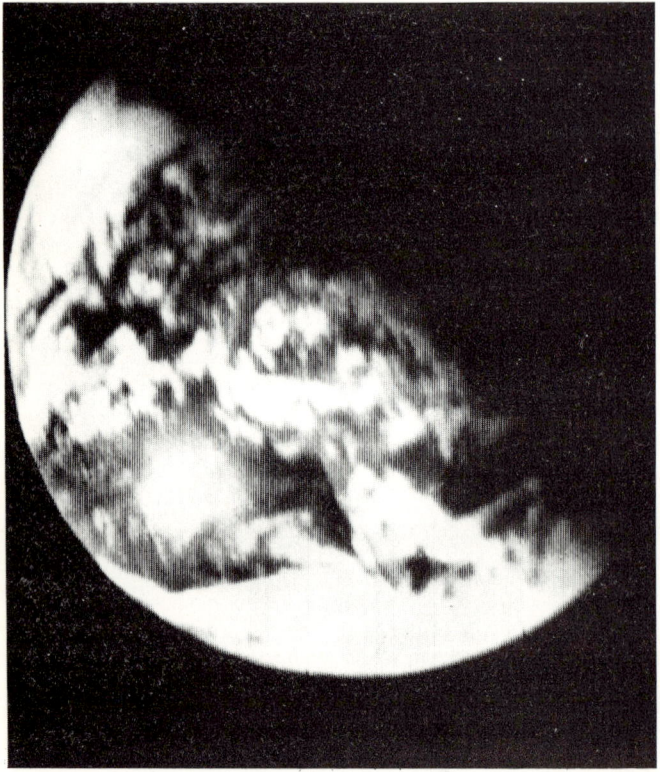

Fig. 2. The earth, largely covered with clouds, as seen by Frank Borman and his team from the *Apollo 8* spacecraft on December 26, 1968, when half way to the moon. The seas appeared deep blue, the land pinkish brown, and the clouds white. The South Pole is at about the 5-o'clock position, and part of South America is recognizable on the lower left. (*U.S.I.S.*)

The Planet Earth

peaks, and penetrates for a few miles into the lithosphere in some regions. On the average, it would be represented by less than the thickness of a sheet of paper on the pile of 3,400 sheets. The study of the biosphere, as such, touches geophysics only in such matters as the influence of vegetation on climate and on the composition of the atmosphere, and the formation of rocks by organisms or from organic remains. (See *Appendix*.)

The diameter of the earth in Fig. 1 is 26 inches, and it is evident that its surface is relatively very smooth. In fact, it is much smoother than the surface of an orange, for example, and is perhaps as smooth as a pearl. Seen from the appropriate distance it might look remarkably like a pearl, in some circumstances, but in 1862 William Thomson (later Lord Kelvin) showed that a much better model would be a steel ball, which the earth as a whole resembles in rigidity, elasticity, magnetic properties and rotundity. The earth would, however, be only three-quarters as heavy, and its surface would be less dense than its interior. Nevertheless, a steel ball makes a vastly better comparison than the traditional 'apple'!

The appearance of the earth from about 111,550 miles (nearly half way to the moon) is shown in Fig. 2, which is a picture televised from the *Apollo 8* spacecraft at Christmas, 1968. From the moon, the earth would appear the size of a ball one inch in diameter as seen from a distance of nearly a yard. The earth's motions would not be directly perceptible from this distance (240,000 miles), but at intervals of an hour or so slight changes due to the earth's rotation might be noted. To appreciate the movement of the earth along its orbit, from this distance, an observer would need to be able to see the movement of the minute hand of a wrist watch.

The Shape of the Earth

The earth was presumed to be a perfect sphere from the

earliest classical times until the 17th century, when René Descartes thought it more likely to be an ellipsoid or oblate spheroid (i.e. 'flattened' at the poles). If the earth were ever in a molten state its rotation would cause the equator to bulge by centrifugal force, and on this supposition both Isaac Newton and Christiaan Huygens calculated the degree of 'polar compression'. Newton's estimate turned out to be a little too small and Huygens' a good deal too large, but in 1743 a very nearly accurate figure was arrived at by A. C. Clairaut from a purely abstract study of the equilibrium of a rotating fluid mass.

Clairaut had been assistant to P. L. M. de Maupertuis when, in 1736, he measured a degree of latitude in Lapland and compared it with a degree at the latitude of Paris, thus demonstrating that the polar compression is indeed a fact. However, his measurements were very far from accurate and the value of the compression was not certainly known to a close approximation until a second expedition to Lapland was made in 1801-1803. The results have been refined more recently by methods which include measuring small variations in the moon's motion, the exact tracing of the precession of the earth's axis (see note on p. 42) and the measurement of the force of gravity at different places all over the world by means of pendulums and spring balances. Combining all these methods, Harold Jeffreys, in 1948, arrived at a polar compression of $1/297 \cdot 1$, but observations of the motions of artificial satellites have since amended this to $1/298 \cdot 2$. This means that the polar diameter of the earth is about $26\frac{1}{2}$ miles shorter than the equatorial diameter.

This does not imply that the earth is a perfect oblate spheroid, for the continents project above the mean surface and the ocean floors descend below it. These modifications show peculiarities of their own which led Lowthian Green to suggest a remarkable theory of the shape of the earth in 1875. He was struck by the triangular shapes of North America, South America, India and, to a less

The Planet Earth

degree, Africa, and explained them by what became known as the 'tetrahedral hypothesis'.

The elevations and depressions on the earth's surface were at that time generally considered to be the results of shrinkage in the cooling earth. Mountains and valleys arose because the solid crust covered a shrinking interior, thus becoming wrinkled like the skin of a shriveled apple. Lowthian Green set out to determine what sort of distribution such wrinkles would be likely to adopt.

He thought, probably correctly, that when the crust was first formed it would show the figure in which the greatest possible volume is contained in a surface of the *least* possible area, that is, a sphere or something very near it. As the interior of the earth shrank, the spherical crust would collapse onto it, tending to produce the figure in which a given volume is contained in a surface of the *greatest* possible area, namely a tetrahedron (or triangular pyramid). If the coigns (or projecting angles) of the tetrahedron were presumed to stand out as the continents, they would indeed form triangles with their angles disposed more or less as they appear on the globe.

This ingenious theory was thought likely enough to be true until the mechanics involved were closely examined. It then became evident that although a very small body might behave in this way, the earth is so massive that a tetrahedron could never form. The weight of the coigns would be so great that they would collapse into the crust, which would again assume the stable form approaching a sphere. The triangular shapes of the continents are evidently an accident after all.

Another theory of considerable interest purported to account for the Pacific Ocean as a great scar left on one side of the earth by the removal of enough rock material to form the moon. According to Osmond Fisher, this material was torn out by the tidal action of the sun during the birth throes of the solar system, when the earth itself was scarcely formed. Thus, he thought, the moon was literally born of the earth, much as the earth was thought (until recently) to be born of the sun by the attraction of a passing star. Fisher's theory was abandoned when it was found that the tidal energy required would not have had this effect, but would have split the earth in half. It was also pointed out that the volume of the moon is about 37 times the volume of the Pacific Ocean!

Clearly, the shape of the earth does not conform to any ideal figure, but it is no easy matter to determine exactly how and where it departs from a perfect sphere, except for the large-scale feature of the polar compression. However, the advent of artificial satellites in 1957 made refined measurements possible, and special *geodetic* satellites have been put into orbit for this purpose. The first of these, *Anna 1 B*, was launched in 1962 and it carried lights that could be flashed at intervals as it crossed the sky. The simultaneous observation of these lights from two or more tracking stations enabled the height of the satellite to be calculated to within a few feet; this was then compared with the height it should have if the earth were a perfect sphere.

By 1968 a number of geodetic satellites had been put into orbit, including *Geos* and *Pageos*, which reflect the sunlight, and *Largos*, which is illuminated by a ground-based laser beam. Detailed analysis of the results obtained has shown that the earth does, in fact, have four slight bulges whose centers lie, respectively, in western Europe, the Pacific south of Japan, Bolivia, and the Southern Ocean near Crozet Island (south of Madagascar).

These bulges have nothing to do with the elevation of the land above sea level, but are part of the mean figure of the earth, which is called the *geoid*. They rise about 220 feet above the mean sphere, while the lowest areas between them drop about 250 feet below it. Further, the equator itself is not circular but slightly egg-shaped, the 'pointed' end lying to the north of New Guinea. There is a bulge over the North Pole but a depression over the South Pole, both far too slight to affect the total polar compression.

The Origin of the Earth

The origin of the earth is still unknown, but the old theories that assumed it to have been 'born' out of the sun have been shown to be untenable. Not only is it impossible to find a plausible process, but the chemical constitution of the sun, which consists chiefly of hydrogen, is quite unlike that of the earth. Current theories maintain that the sun gathered about itself large quantities of dust and gases, either in its own parent nebula or in passing through one of the clouds of such matter which occur in the galaxy, and that rotation distributed this matter round the sun in the form of a disk. Turbulence in the disk brought enough of the solid particles sufficiently close together to start a sort of condensation by gravity. Once a large enough globe of matter had gathered, it would continue to grow as other particles fell onto it, and eventually a body the size of the earth resulted. Even now, it is estimated that about 8,000 million meteors enter the earth's atmosphere every day, and some ten tons of solid matter from space land on its surface.

The origin of the material in the primeval dust cloud is anybody's guess, but the presence of the heavier elements suggests that much of it was produced somewhere from intense nuclear activity, either during the early history of the galaxy itself, or perhaps in the heart of the type of exploding star called a *supernova*. The original elements must have included a great many radioactive atoms which have since decayed into stable ones, but radioactive elements probably played an important part in the second chapter of the earth's history. The initial condensation would have resulted in a considerable rise in temperature, and this, augmented by the radioactive heat, probably rose above the melting point of all the elements fairly soon after the earth's formation. The earth then became a ball of molten rock which has cooled down, on the whole, ever since, though with periods of temporary reheating which we shall describe in due course.

One of the earliest attempts to describe the earth's structure on scientific lines was made by René Descartes in 1640. He proposed that the earth had once been a small sun, and that remnants of its incandescent metallic vapors still occupy its center. This is covered by a layer of a dark material of which he thought sunspots consist, and this again by a shell of heavy rock. Outside this lay an ocean enclosed, in its turn, by a shell of lighter rock broken into large fragments. Some of these were tipped up so as to project and form mountains and continents, while water flowed through the cracks over other parts to provide seas on the surface. The water circulated between the upper and nether seas like 'the blood in the bodies of animals'. A better and simpler model was proposed by Gottfried von Leibniz later in the same century. The eruption of white-hot lava from volcanoes led him to the view that the earth is a ball of molten rock whose surface layers have cooled to form a comparatively thin solid crust. This view was generally accepted for a very long time, and the term 'crust' is still applied to the outermost layer of rock, though we now know that it rests on more solid rock of much greater thickness.

The Earth's Heat

We have no direct knowledge of the earth below the depth to which mines and wells have penetrated. The deepest mine* descends 11,246 feet (2·13 miles), and the deepest boring† 25,340 feet (4·8 miles). Since the highest mountain

*The East Rand Proprietary Mine, Boksburg, Transvaal. The gold workings begin nearly 2 miles down, and the maximum depth was reached in 1959.
†A trial bore-hole made by the Phillips Petroleum Company in Pecos County, Texas, in 1958. No oil was found.

(Mt. Everest) rises 5·5 miles above mean sea level, and the deepest part of the ocean (Marianas Trench) descends 6·85 miles below it, the mining and boring achievements are notable as feats of engineering rather than as significant probes toward the earth's center, which lies 3,950 miles farther down. Yet they have provided useful information about the earth's heat.

It was observed long ago that the temperature of the earth in a mine increases with the depth, the rate of increase being termed the *geothermal gradient*. This varies greatly from place to place and is naturally high in the neighborhood of volcanoes. In the regions where early measurements were made it averaged 1°C. for every 115 feet (or 1°F. every 64 feet), but it is known to be much less than this in places remote from volcanic activity. The theoretical gradient averages about 1°C. per 355 feet through the first 30 miles, and then falls off rapidly until 50 miles is reached, where it is only 1°C. per 4,000 feet and the temperature averages about 500°C. Rocks are very poor conductors of heat, and at this level the overlying layers act like a thick blanket and virtually prevent the heat from escaping.

The rate at which a substance conducts heat is known as its *thermal conductivity*, and is measured as the number of calories passing through one cubic centimeter in one second, if a temperature difference of 1°C. is maintained between the opposite faces. If its value for the crustal rocks is multiplied by the geothermal gradient, we get the actual amount of heat passing upwards through the crust of the earth per second. This is called the *heat flow*, and it turns out to be very small. The average for the world is about 37,000 calories per square foot of the surface per year, and it would take five years to melt a sheet of ice one inch thick, or one year to melt as much as ordinary sunshine will in a single afternoon.

However, the heat-flow varies in different places. In Great Britain, for example, it is slightly below the average, but in the hot-springs region of central Japan it is as high as 354,000 calories, and here the rocks must be very near their melting point (about 1,000°C.) only a few miles down. The heat flow in the ocean floor was measured off the California coast in 1954 and was found to be very near the average for the continents, but a rather lower value was found in the eastern Atlantic by the research ship *Discovery II*. A much higher heat-flow was detected in mid-Atlantic on the Dolphin Ridge, and in the East Pacific Rise, but exceptionally low values turned up again in the ocean deeps. These results have an important bearing on events believed to be taking place deep in the interior of the earth, as will be shown in the following chapter.

The earth is thus continually losing heat, which is ultimately radiated away into space, but is it a simple case of a hot body slowly cooling down? For a long time this was regarded as so obvious that attempts were made to estimate the age of the earth from its present temperature. In 1862 William Thomson (Lord Kelvin) calculated the earth's present store of heat from the geothermal gradient, and came to the conclusion that the earth's crust must have been molten only 100 million years ago. This greatly upset Charles Darwin, who wanted at least ten times as long for his theory of evolution, but others made similar calculations and in 1899 Archibald Giekie thought 100 million years adequate for the deposition of the sedimentary rocks. However, new geological evidence was already turning up which demanded much more time, and by the 1920's some 2,000 or 3,000 million years were thought to be necessary. More recently the oldest known rocks have been dated by the radioactive minerals they contain. Igneous rocks containing uranium-238 are often used for this purpose. Uranium-238 disintegrates through 17 stages, finally becoming lead-206, which may be distinguished from the ordinary lead-207. Every 20,000 atoms of the uranium present in the rock when it

The Planet Earth

first cooled to a solid produce 10 atoms of lead-206 every million years (this rate may be considered steady over a few thousand million years, because uranium-238 has the enormous 'half-life' of 40,000 million years). Thus, the age of the rock can be calculated from the proportion of lead-206 to uranium now found in it. Other radioactive elements, such as thorium (which changes to lead-208), can be used in the same way. This method of dating has set back the solidification of the earth's crust to a minimum of 4,500 million years ago. Yet the earth is very far from being a cold body, and it is evident that it must have its own internal source of heat.

There is only one possible source known to science, and that is radioactivity, but the reactions are not of the kinds that occur in atomic piles and bombs, or in the sun. The heat generated in the earth is the normal accompaniment of radioactive disintegration, and comes primarily from scattered atoms of thorium and uranium, and their transmutation products (which include radium and polonium). One gram of radium releases the heat of a half-burnt match every hour, and though uranium requires 30 million times as long (about 3,400 years) to yield the same amount of heat, it is also about 30 million times as abundant.

In 1906 Robert John Strutt (Lord Rayleigh) discovered minute traces of radioactive elements in a large number of common rocks gathered from all over the world. He showed that there is more than enough in the crustal rocks to maintain the earth's temperature at the observed level. The heat is generated faster than it escapes so that the earth must be very hot indeed in the central regions. Far from cooling, it is actually getting hotter at the present time.

Chemical Composition

Chemically, the earth appears to be fairly typical of the

smaller planets, though we do not know of another with liquid surface water or an atmosphere rich in oxygen. The first 94 elements in order of weight (from hydrogen to plutonium) have all been found in the earth's crust. However, a few (such as the radioactive astatine, neptunium and plutonium) occur only in spectroscopic traces, generally in uranium ores. There are substantial quantities of about 70 elements, most of which are metals and all of which have been found in the sun, though in quite different proportions. The general proportions of the elements in the earth's crust are shown in Table II.

This table includes a great many elements commonly considered extremely rare, but in fact considerably more abundant than silver or gold. They are scarce only because their ores do not occur in economically workable concentrations, but they are increasing in importance as their special properties find uses in technology. Europium, for example, for long no more than a rare curiosity, is now in demand for the red phosphor in color television, and terbium, first isolated about 1940, is required for use in lasers, transistors and other electronic devices.

Table II also shows the abundance of the elements in terms of the proportion of their atoms, the different result being due to the relative weights of the atoms. For example, there are 9 elements more common than hydrogen by weight, but only 3 by number of atoms. Again, beryllium is nearly twice as rare as tungsten by weight, yet there are nearly twice as many beryllium atoms in the earth's crust as tungsten atoms. Helium and gold are almost equally rare by weight, but for every atom of gold there are 30 of helium; titanium is nearly 4 times as common as hydrogen by weight, yet its atoms are 15 times as rare. It should be noted that the order of the elements in terms of the number of atoms does not give the order in terms of volume, for though all atoms are roughly of the same order of size there is great variation in the densities of the elements in bulk. (For example,

Table II: THE EARTH'S CRUST

W: Parts per 1,000 million of the elements, by weight.
A: Atoms per million million. m = million. * = non-metal.

	W	A		W	A
*Oxygen	467 m.	29,000 m.	Germanium	7,000	96,000
*Silicon	277 m.	9,900 m.	Samarium	6,500	43,000
Aluminum	81 m.	3,000 m.	Gadolinium	6,400	41,000
Iron	50 m.	900 m.	Beryllium	6,000	700,000
Calcium	36 m.	907 m.	Praseodymium	5,500	39,000
Sodium	28 m.	1,220 m.	Scandium	5,000	100,000
Potassium	26 m.	662 m.	*Arsenic	5,000	70,000
Magnesium	21 m.	871 m.	Hafnium	4,500	25,000
Titanium	4·4 m.	92 m.	Dysprosium	4,500	28,000
*Hydrogen	1·4 m.	1,400 m.	Uranium	4,000	17,000
*Phosphorus	1·2 m.	38 m.	*Boron	3,000	273,000
Manganese	1 m.	18 m.	Ytterbium	2,700	15,600
*Sulphur	520,000	16 m.	Erbium	2,500	15,000
*Carbon	320,000	27 m.	Tantalum	2,100	11,600
*Chlorine	314,000	9 m.	*Bromine	1,600	20,000
Rubidium	310,000	3·6 m.	Holmium	1,200	7,270
*Fluorine	300,000	16 m.	Europium	1,100	7,200
Strontium	300,000	3·4 m.	Antimony	1,000	8,200
Barium	250,000	1·8 m.	Terbium	900	5,660
Zirconium	220,000	2·4 m.	Lutetium	800	4,570
Chromium	200,000	3·8 m.	Thallium	600	2,900
Vanadium	150,000	3 m.	Mercury	500	2,487
Zinc	132,000	2 m.	*Iodine	300	2,360
Nickel	80,000	1·4 m.	Bismuth	200	957
Copper	70,000	1·1 m.	Thulium	200	1,190
Tungsten	69,000	375,000	Cadmium	150	1,330
Lithium	65,000	9·3 m.	Silver	110	1,020
*Nitrogen	46,000	3·3 m.	Indium	100	870
Cerium	45,000	320,000	*Selenium	90	1,140
Tin	40,000	336,000	*Argon	40	1,000
Yttrium	28,000	314,600	Palladium	10	93·5
Neodymium	24,000	167,000	Platinum	5	25·6
Niobium	24,000	258,000	Gold	5	25·4
Cobalt	23,000	400,000	*Helium	3	750
Lanthanum	18,000	130,000	*Tellurium	2	15·7
Lead	16,000	77,300	Rhodium	1	9·7
Gallium	15,000	214,000	Rhenium	1	5·4
Molybdenum	15,000	156,000	Iridium	1	5·2
Thorium	12,000	51,700	Osmium	1	5·3
Cesium	8,000	60,000	Ruthenium	1	9·9

the densest element, osmium, is twice as heavy as lead, yet its atom is lighter than the lead atom.)

There are 14 elements (including radium and polonium) too rare to list in Table II, but even the rarest mentioned – ruthenium, of which there are only 10 atoms in every million million – has an industrial use as a catalyst and a hardener of alloys. It costs about twice as much as gold, but rarity is sometimes less important than the difficulty of extraction. For example, europium, which is more than 1,000 times as abundant as ruthenium and 220 times as abundant as gold, costs about five times as much as gold. Elements that occur in mere traces are vastly more expensive. Radium is priced at about $34 per milligram which is equivalent to about $956,000 per troy ounce or around 40,000 times the price of gold! About 16,000 tons of ore would be required to produce one troy ounce of radium, yet it has been estimated that there are some 150 million tons of radium in the earth's crust – scattered about in microscopic specks.

3

The Lithosphere

The study of the lithosphere naturally began with the geology of the earth's outer crust, and one of the first geophysical problems was the formation of mountains. It was known from ancient times that rocks at great heights above sea-level sometimes contain fossil sea-shells, and this seemed to be explained by the even older legends of a primeval Flood, in which all the land was covered by water in some remote cataclysm. But it appeared doubtful that there could be enough water on the earth to cover all the mountains simultaneously.

It was known that the level of the sea has changed from time to time, for buildings erected on land were found centuries later to be standing in the sea, and ancient harbors with moorings for boats were sometimes found high and dry some distance inland. Since such changes were not always of the same kind at places along the same coast, some of them had to be attributed to local elevations or depressions of the land. It remained for Leonardo da Vinci (1452–1519) to show that elevation of the land was the main cause of the occurrence of sea-shells on mountain tops, and that they could not be accounted for by a universal Flood.

Mountain Building

At first, all high land was presumed to have been raised by vertical movements, but in 1846 Henry De la Beche showed that while this could be true in cases where the strata remained very nearly level, it could not account for *fold* mountains, in which the strata were sometimes folded right over and even doubled back on themselves. Mountains of this kind must have been produced by horizontal movements of the earth's crust, which buckled up the areas of less rigid rock just as your foot will buckle up a rug if you try to push it along the floor.

Of course this works best if you try to push the rug against the wall, and in 1873 the American geologist James Dwight Dana thought that the continents form rigid, stable blocks which would act like the wall if the level sea floor were pushed against them. He observed that the Rocky Mountains and the Andes appeared to have been buckled upwards in exactly this way. Since they contain great thicknesses of sedimentary rocks, originally deposited in level layers on the bottom of the sea, their present site must once have been occupied by a vast depression in the ocean floor. Such depressions he called *geosynclines*.

After an intensive study of the Alps, Eduard Suess and Albert Heim came to the conclusion that the Alps had been formed in the same manner from a geosyncline in a large sea running from west to east across southern Europe. Suess called this sea the 'Tethys,' and its extension eastwards across Asia provided for the deposition of the rocks which now form the Himalayas. The squeezing movements which raised all these mountains must have constituted a major event in the earth's history, and he called it the 'Alpine orogenesis' (which means 'Alpine mountain building'). It took place in Tertiary times about 50 million years ago and may still not quite have died away.

Suess also recognized older periods of intense mountain-building, such as that responsible for the upheaval of Pre-Cambrian rocks in Scotland 400 million years ago. This he named the 'Caledonian orogenesis' (from *Cale-*

Table III: THE LAST SIX REVOLUTIONS

Years Ago (millions)	America	Europe	Elsewhere
50	Pasadenian Laramide Andean	Alpine	Alpine (Asia)
250	Appalachian	Hercynian	Cape Folds (Africa) Tasman (Australia)
450	Acadian Taconic	Caledonian	Early Tasman (Australia)
650	Wichita	Early Caledonian	Late Katanga (Africa) Adelaide (Australia)
850	Killarnean	Charnian	Delhi (India)
1,050	Grenville	Gothic	Satpura (India) Gordonian (Africa) Musgrave (Australia)

Note: Each 'date' refers broadly to a period of several million years, which either contained it or fell close to it. The phases of activity often occurred in different parts of the world at somewhat different times, and the lack of exact correlation justifies the use of the different regional names.

donia, Scotland). Some 18 such *revolutions*, as they are sometimes called, have since been discovered. They have been dated by the radioactive minerals in the igneous rocks associated with them, and are found to occur, on the average, at intervals of about 200 million years. Each revolution occupies from 20 to 50 or more million years for its completion, and the oldest known took place 3,400 million years ago. The last six are given in Table III.

During the first quarter of the present century the Irish physicist and geologist, John Joly, and others realized that the radioactive heat accumulating in the deeper rocks of the earth must eventually rise to near their melting-point. At such a time the rocks at no great depth would become plastic if not liquid, and the movements required for mountain-building on a major scale would be enormously speeded up. At such times the heat flow from the interior would greatly increase, and the rapid escape of heat would continue until the rocks below the surface had cooled sufficiently to become solid again. Another period of heating up would then begin and in due course the cycle would be repeated. The time required for this to happen turned out to be about 200 million years, so that the occurrence of the revolutions could in this way be accounted for.

Floating Continents

The world's great mountain ranges rise from continents which already stand higher still above the level of the sea floor, and as long ago as 1859 George Airy considered that the earth's crust could not possibly support their weight. They must, he thought, consist of lighter materials than the crust and have been floating in it when the crust first solidified. This would leave them in a state of equilibrium, with deep supporting 'roots' extending downward beneath the highest land. At about the same time Archdeacon J. H. Pratt drew attention to the fact that in experiments with pendulums the large mountain ranges have been shown to be deficient in gravitational attraction, and this could only mean that they are composed mainly of comparatively light rocks.

Igneous rocks had already been found to fall into two broad classes differing in density. Those consisting mainly of silica and aluminium, such as granite, form a lighter group than those composed mainly of silica and

The Lithosphere

calcium or sodium, such as basalt, which are also found to be rich in magnesium and iron. Suess proposed referring to the first type of material as *sal*, now amended to *sial*, from the first letters of *si*lica and *al*uminium, and to the second and heavier type as *sima* from *si*lica and *ma*gnesium.

It seemed to C. E. Dutton in 1889 that the cores of the continents consist mainly of sial, and that they float in the manner described by George Airy in a crust composed chiefly of sima. He called this principle *isostasy*, and worked out the consequences in considerable detail, calculating that the 'roots' beneath the highest mountains probably extend downwards for at least 25 miles. See Fig. 3.

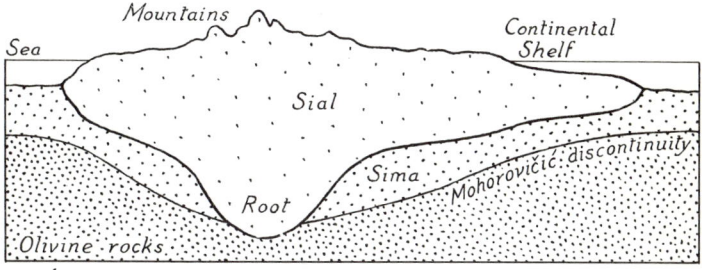

Fig. 3. The principle of isostasy. The continental block of sial floats like an iceberg in the denser sima.

When the earth was molten, the heavier rocks would naturally gravitate downwards and the lighter rocks float on top. Suess proposed that the crust of the earth consists essentially of a shell of sima, covered by a thinner shell of sial which is gathered into thicker masses on the sites of the continents. Below the sima there must be still heavier rocks, such as peridotites or olivine rocks, which are known (from intrusions exposed on the surface) to consist chiefly of the silicates of iron and magnesium. If this were all, the density of the earth as a whole would still be much too low, so he also postulated a large central core of nickel and iron, calling this material *nife*, from *ni*ckel and *f*errum (iron). The earth's magnetism had long been thought to be evidence of an iron core, so that Suess was able to construct a plausible model of the earth which would have the known density of 5·5.

A density of 5·5 means that the earth is 5·5 times as heavy as water, and this was first ascertained in 1798 by Henry Cavendish, who arrived at the approximate figure of 5·48. His classical geophysical experiment consisted in measuring the movement of two small lead balls by the gravitational attraction of two large ones, in a torsion balance. The time required for the movement was also recorded, and the masses and distances apart of the balls being known, these quantities could be expressed as a force. The density of the earth was then found by comparing this force with that exerted by the earth in producing the well-known acceleration of falling bodies.

Suess not only accounted for the earth's density, but he also attempted a picture of the ancient geography of the globe. The similarities between some geological features at corresponding latitudes on opposite sides of the Atlantic Ocean, and between South Africa and Australia, led him to believe that the North Atlantic was at one time above sea level and formed a land bridge between North America and Europe, and that similar land bridges connected South America, Africa, Australia, and peninsular India. There were thus only two gigantic continents, a double one in the northern hemisphere consisting of 'Arctis' (from North America to Europe) joined to 'Angara Land' (northern Asia), and a quadruple one mainly in the southern hemisphere, which Suess named 'Gondwanaland.' They were separated by the Tethys (page 34), evidence of the separation being afforded by the distinctive plants and animals, both extinct and living, of the two hemispheres.

The typical plants of the coal forests of the southern hemisphere were quite unrelated to those of the northern coal forests, and even now the animals are largely of different types. For example, there are about 256 species of marsupials living in the Gondwanaland group of continents, but only 2 in the northern lands.

Continental Drift

Nevertheless, Suess' land bridges, which had to be 3,000 miles or more long, were something of a tall order, and oceanographic surveys showed no evidence of any sial on the ocean floors, where the bedrock seems to be wholly of sima. A more plausible theory, from this point of view, was that of *continental drift*.

This theory maintains that the continents are detached blocks of sial 'floating,' like icebergs, in the outer crust of the earth, which is of sima. Like icebergs, too, they can drift about, and those that are now far apart may once have been close together, making land bridges unnecessary to explain their similarities. That the continents have not always occupied their present positions was first proposed in 1620 by Francis Bacon, who noted that the east and west coasts of the Atlantic Ocean would fit together like a jig-saw puzzzle if America were pushed across to join Europe and Africa. A similar observation was made in 1668 by the Frenchman P. Placet, who thought that 'America was not separated from the other parts of the world' before the time of the Flood. The subject was raised again in 1855 by the Belgian Antonio Snider, who considered on the evidence of fossil plants that the Americas broke away in late Carboniferous times and drifted slowly westwards to their present positions.

The cold crustal rocks beneath the oceans are, of course, solid, but that the continental blocks could move through them was shown to be possible by experiments performed in 1870 by Auguste Daubrée. He discovered that rocks subjected to very great pressure, sufficiently slowly, would yield without fracturing like highly viscous fluids. There was also abundant evidence of the plastic nature of rocks in the folded strata of anticlines and synclines. During the driving of the St. Gotthard Tunnel in Switzerland, from 1872 to 1880, a row of crowbars let into the rock were observed to lose their exact alignment as the rock slowly yielded without fracturing to pressures from the mountain.

There was, however, no known force sufficiently great to move the continental blocks, though in 1908 F. B. Taylor, who was chiefly concerned to account for the great mountain ranges, thought that the moon was once close enough to the earth to exert an adequate pull. That this could not have happened was shown by Harold Jeffreys, who calculated that if the moon could have exerted enough force to move the continents, the tidal friction generated would have brought the rotating earth to a standstill within a year!

Nobody had any better suggestion to offer and the whole idea of continental drift began to seem improbable. Then, in 1912, the German Alfred Wegener published an impressive mass of new evidence in its favor. He elaborated the correspondence on both sides of the Atlantic in detail, and showed from an examination of ancient climates that the continents were once differently situated with regard to the poles. In both Carboniferous and Eocene times, Europe supported tropical forms of life. In Permain times, there was an ice age in the southern hemisphere which seemed to place the South Pole near the coast of South Africa, while at the same time luxuriant coal forests flourished in Antarctica.

Wegener proposed that originally all the great landmasses of the world, including Antarctica, were packed together to form a single giant continent, which he called 'Pangaea.' See Fig. 4. This spread practically from pole to pole, and was broken up in late Carboniferous times by the gravitational pull of the earth's equatorial bulge on the polar areas. The detached pieces engaged in a

The Lithosphere

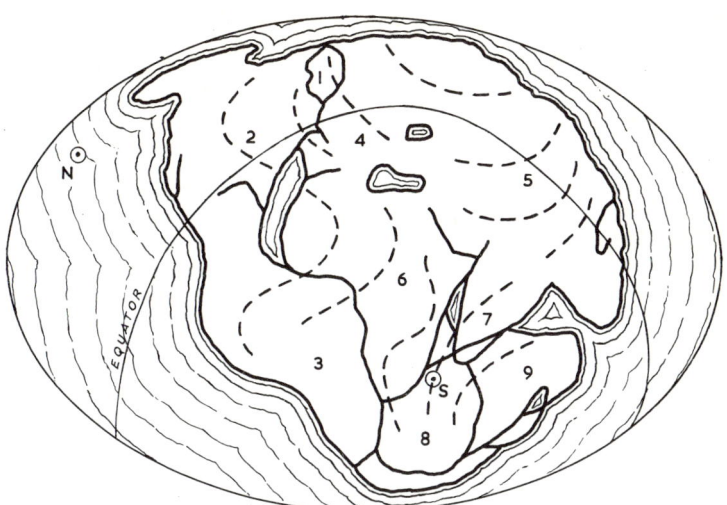

Fig. 4. Wegener's *Pangaea*, showing probable positions of the equator and poles in Carboniferous times, about 350 million years ago. 1, Greenland. 2, North America. 3, South America. 4, Europe. 5, Asia. 6, Africa. 7, India. 8, Antartica. 9, Australia. N, North Pole. S, South Pole. The dotted lines show the trends of mountain ranges folded about 3,000 million years ago.

'flight from the poles,' and under the influence of the sun and moon sought a balanced position round the equator. Their movement against the resistance of the earth's crust squeezed up high mountain ranges along their leading edges, while detached fragments were left in their wake as festoons or 'arcs' of islands. The Rockies and Andes are examples of such fore-edge folding, and the West Indies, the Kurile Islands and Japan form typical festoons.

Wegener thought it might be possible to measure directly the annual westward motion of North America, and particularly of Greenland, where he claimed to have detected a movement of 20 yards per year. However, his data were unreliable and not confirmed by independent observation, and no undoubted motion has yet been directly measured. Further, the gravitational forces invoked by Wegener were then shown to be about 10,000 million times too weak! It was also pointed out that the evidence of ancient climates could equally well be explained by what was called the 'wandering of the poles.'

If the South Pole once lay off Africa, and Paris once stood on the equator, this could mean simply that the North and South Poles have themselves shifted. However, it is inconceivable that the axis of the spinning earth could change its direction in this manner* so that the 'wandering of the poles' really means the wandering of the entire outer shell of the earth, *en bloc*, over its interior. It now seems certain that this has really happened. Repeated observations of star positions at four stations on the 40th parallel of latitude (in Europe, America and Japan) indicate that the earth's crust has shifted in this manner about 30 feet during the last 60 years. The position of the rotational pole remains unchanged, but its observed position on the surface, where the meridians of longitude meet, thus appears to wander over the map. However, this sort of motion leaves the relative positions of the continents unchanged and all but the climatic evidence for continental drift unexplained. Wegener realized this and began to look for a more adequate source of power.

In 1928 he revived an idea dating back to 1839, when William Hopkins suggested that the molten rock then believed to lie beneath the crust of the earth probably circulates in convection currents, just as the water in a kettle over the fire circulates. In 1881 Osmond Fisher considered that such currents traveling along beneath

---

*The movements demanded would be quite different from a simple 'wobbling' of the earth's axis, which would merely cause the poles to describe small circles. Three such wobbles are known, the smallest being the 'Chandler wobble,' in which the axis rotates about a mean position (known as the 'secular' pole) at a distance of approximately 20 feet once every 14 months. The other two wobbles are the *nutation*, a 'nodding' of the axis every $18\frac{1}{2}$ years, due to the moon, and the larger wobble which causes the precession of the equinoxes and has a period of 26,000 years.

The Lithosphere

the surface would drag the overlying crust and so provide the horizontal movements required for buckling up the mountain ranges. This idea of convection currents had been dropped when the study of earthquakes showed that the rocks beneath the crust are not liquid but solid, and it seems to have been forgotten when Daubrée later demonstrated that solid rocks also may 'flow.' However, in 1928 Harold Jeffreys found that convection currents in the solid rock are quite possible if the rocks are very much hotter underneath than on top. Wegener saw that such currents, even if they move only about 1 inch per year, could move the continental blocks along quite fast enough to explain continental drift.

This idea was fully worked out by A. L. du Toit in 1937, but his theory differed from Wegener's in supposing that there were originally two great continents instead of only one. In the northern hemisphere lay 'Laurasia,' consisting of North America and Eurasia packed together. In the southern hemisphere, South America, Africa, peninsular India, Australia and Antarctica were similarly fitted together to form 'Gondwanaland.' The sub crustal currents, pushing against the roots beneath these landmasses, eventually moved them to their present positions. A movement of 1 inch per year would amount to 3,000 miles in 200 million years, but a much slower rate would suffice for most of this time if it were speeded up during the 'revolutions.'

The convection currents, now generally accepted, take place in the deeper rocks beneath the crust, a zone known as the *mantle*, but the crust itself takes part in the surface movements. At some places there must be ascending currents of hot rock which, when they reach the underside of the crust, spread out in opposite directions and travel horizontally. Being near the surface they cool off as they move, and eventually become cold enough to sink once more into the depths. See Fig. 5. The rising currents tend to produce bulges in the crust over them, and increase

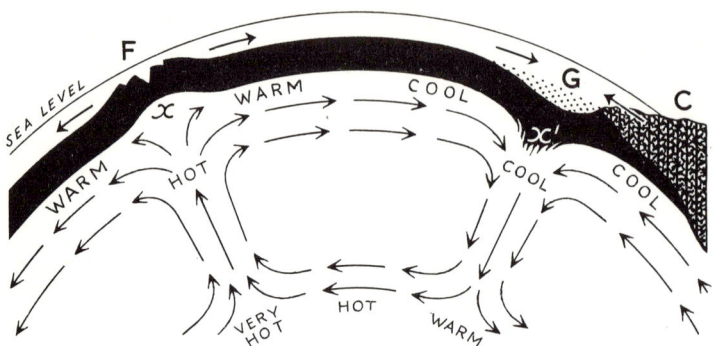

Fig. 5 Convection currents in the earth's upper mantle. At F, the rising currents produce highly faulted mid-ocean ridges and rifts. At x, the rocks are under tension and the pressure is reduced: magma chambers with volcanic outlets may form. At x^1, the descending currents draw down the crustal rocks (solid black) to produce the ocean deeps; reduced pressure from suction may also give rise to magma chambers with associated volcanoes. A continental block, C, drawn toward the depression, supplies the sediments forming the geosyncline G. The rocks here are under compression; intense folding and faulting will eventually occur and mountains will be squeezed up.

the local heat flow. When the cold rocks begin to sink they tend to cause depressions in the crust, which is sucked downward, and there is a low rate of heat flow. Regions showing these characteristics have been found and some are mentioned on pp. 27–28.

The 'floating' continental blocks are carried away from the localities of rising currents, but drawn towards the 'sinks' where the descending currents begin. These sinks are the geosynclines, and as a continent approaches one, the products of its erosion supply sediments to fill the depression at least as fast as it subsides. As the continent approaches still closer it crumples up the layers of sediments to form the 'fore-edge' mountain ranges. That this mechanism would 'work' was demonstrated with models by David Griggs in 1939.

Another consequence of the rising currents follows if

The Lithosphere

they happen to meet the surface beneath the middle of a continent. First, the upward bulge 'cracks' the continental block in half, and then the diverging surface currents tear the fragments apart and carry them off in opposite directions. This is how the original super continents are thought to have been broken up and the pieces dragged off to their present positions. Evidence that something of the sort is going on now is seen in the Great Rift Valley, which splits Africa from south to north and is continued in the Red Sea.

Even after this theory had been plausibly worked out, many geologists refused to believe that it is really true. One of the principal objections was that most of the continents must have rotated as well as moved along in order to attain their present positions. This seemed to be contradicted by the occurrence of fossils of tropical, temperate and Arctic species in parallel zones which are still parallel with the equator. This evidence is fragmentary, however, and is offset by other fossil evidence of extensive climatic changes demanding at least a 'wandering of the poles' on a large scale. To settle the matter, independent evidence of the rotation of the continents was therefore sought.

This was found in 1959 and 1960 by P. M. S. Blackett and others, who examined igneous rocks containing magnetic minerals gathered from all over the world and analyzed the results. Such minerals* are often constituents of the lava thrown out by volcanoes, but they are not magnetic in the molten state. As the lava cools and the minerals crystallize, they become permanently magnetized with their poles aligned with the magnetic poles of the earth, like tiny fossil compass needles. They then show the north and south directions at the place where they were formed when the eruption occurred. This may have been

---

*The most common is magnetite, or black oxide of iron (Fe_3O_4), the substance of which 'lodestone' is composed.

many millions of years ago but it can usually be placed in the correct geological period.

Many of the specimens examined had evidently been turned to point in quite new directions since they first crystallized, and specimens of different ages from the same continent usually showed progressive changes in direction, so that the continent bearing them must have rotated. The different continents had, moreover, turned in different ways during the same periods, so that the changes in direction could not have been caused by movements of the earth's magnetic poles, which would have produced the same results in all the continents. More evidence of the same kind has been found since 1960, and the fact of continental drift has now been put beyond reasonable doubt.

Further studies of rocks containing magnetic minerals have even enabled the speeds of movement of some of the continents to be estimated. During the quiet periods between revolutions a mean speed of about $\frac{3}{4}$ inch per year seems to be indicated, but in 1966 evidence was found in Pakistan that this part of Gondwanaland speeded up to 6 or even 8 inches per year during the Permain (Hercynian) revolution, but dropped back to the slower speed as the activity died away.

However, the interpretation of the data is by no means simple, and it was further complicated by the discovery in the 1960's that the earth's magnetic poles have themselves changed over or 'reversed' at intervals of about 450,000 years, north becoming south and south north. These changes have been traced back to Cretaceous times, 76 million years ago, the evidence coming from the sea floor. As new rocks arrive at the mid-ocean ridges from deep in the mantle they become magnetized as they cool, and in this state they move off away from the ridges to make room for younger rocks coming up from below. These become magnetized in their turn, and on each side of the ridge there is a succession of rocks showing the

The Lithosphere

state of the earth's magnetism at different periods. The rocks occur in alternate strips with opposite magnetism, showing that the change over of the poles is comparatively sudden. Radioactive dating of rock samples indicates that this 'sea-floor spreading,' as it is called, is taking place at about ¾ inch per year in the South Atlantic, but nearly twice as fast in the North Pacific. (See Fig. 9.)

Has the Earth Expanded?

Wegener found he could fit all the continents together like a jigsaw puzzle to make a single land-mass. In 1965 K. M. Creer found by means of models that he could fit them together so as to cover completely a globe with a diameter about half that of the earth on the same scale. This miniature world was thus entirely covered by a layer of sial, as in Suess' original model of the earth. The 'expanding earth' theory – first proposed as long ago as 1935, by J. K. E. Halm – now maintains that the earth originally had this form and was therefore about one-eighth its present size. It is presumed to have been steadily expanding at an average rate of about 1/25 of an inch per year on its diameter for at least 4,500 million years. In doing so it cracked its skin of sial into pieces that did not themselves expand, but now form the continents.

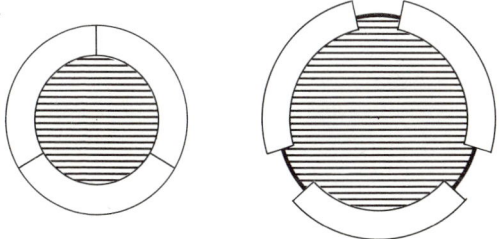

Fig. 6. How the earth's original shell of sial became broken up to form separate continental blocks, according to the expanding earth theory. (Thickness of sial greatly exaggerated.)

There was a short period while they remained together like Wegener's Pangaea, the initial expansion producing only the Pacific Ocean as a new feature. Pangaea then split into Du Toit's Laurasia and Gondwanaland, the subsequent breakup conforming more or less with his theory of continental drift, with modifications to accommodate the expansion. The general principle is illustrated in Fig. 6.

Since the earth is now supposed, on this theory, to have eight times its original volume, it is natural to wonder where all the additional matter came from. Provision for this is made by assuming that the earth was initially eight times as dense as it is now, or about twice as heavy as gold, and has since 'risen' like baker's dough. This is quite conceivable if its temperature and pressure were originally high enough to strip off some of the electrons from the atoms, enabling them to be packed very much closer together. The expansion would then occur as the atoms gradually acquired their full complement of electrons.

A great deal of geological evidence has been found which may at least be explained by the theory of expansion, and it also helps to explain the slowing down of the earth's rotation, for which the orthodox theory of tidal friction is, by itself, scarcely adequate. The earth would certainly expand if the weight of rock squeezing the interior were reduced, and the theory of the expanding universe has been invoked to show that the force of gravity may very well have been steadily weakening with time. But these are all speculations, and the theory of the expanding earth is too new for any definite conclusion to be reached.

The Interior of the Earth

Our knowledge of the interior of the earth is derived chiefly from observation and measurement of the waves or vibrations caused by earthquake shocks. The majority

The Lithosphere

of earthquakes result from the sudden slipping of deep-seated rocks subjected to uneven pressure. Such events usually occur less than about 5 miles below the surface, but while depths of up to 30 miles are not uncommon, shocks below this level are rare. The site of a shock is called its *focus*, and the place on the surface exactly over it is the *epicenter*.

Vibrations radiate in all directions from the focus and may be detected at the surface of the earth by *seismographs*. These sensitive instruments often register earthquake shocks occurring on the other side of the world, so that some of the waves recorded appear to have passed through or near the center of the earth. In 1906 R. D. Oldham showed how the records can be made to reveal the structure of the earth's interior.

Two types of waves are produced: longitudinal waves, which are 'push-and-pull' vibrations like sound waves, or like those passing along a train of railway cars being shunted; and transverse waves, which are 'up-and-down' or 'sideways' shakes, like those that travel along a rope when one end is suddenly jerked. The longitudinal vibrations are called *primary* or *P waves*, and the transverse vibrations *secondary* or *S waves*. Both types of waves will travel through solids, but only the P waves can pass through liquids.

The P waves traveling upward to the surface jolt the epicenter up and down, and this starts another set of transverse waves radiating from the epicenter. These travel over the surface like the circular ripples on a pond into which a stone has been thrown. They are called *Rayleigh waves* because they were predicted by Lord Rayleigh in 1887, twenty years before they were discovered. The S waves traveling upward to the surface start another set of transverse waves radiating from the epicenter, but these are sideways shakes and are called *Love waves* after A. E. H. Love, who first explained them. The Rayleigh and Love waves are collectively known as the *long* or *L waves*.

All these waves travel at different speeds, the P waves being twice as fast as the S waves, which are again faster than the L waves. Further, their speeds vary with the density of the rock, so that the P and S waves travel faster as they go deeper into the earth, where the rocks are under enormous pressure. Since the earth is round, the zones of increasing density are curved and refract the waves like a lens, causing them to bend outwards towards the surface. This means that there is a region on the opposite side of the world to the epicenter where no direct waves are ever recorded. Surrounding this is a circular belt or *shadow zone* in which P waves are recorded but not S waves, showing that a barrier of liquid rock must somewhere have been encountered.

The depth of the focus and the speeds and paths of the waves can be calculated from the times at which they arrive at seismographs situated at different distances from the epicenter, and also from the character of the curves produced. Two new sub types of the P and S waves produced in the depths reveal that the earth consists of three major shells of basically different densities. These have clearly marked boundaries or *discontinuities* from which the waves may be partly reflected. Other details have emerged from the study of the waves produced when H-bombs are exploded underground, and it has now become possible to build up a general picture of the structure of the earth as shown in Fig. 7.

It appears that there is a solid core about 1,700 miles in diameter, but this is no longer believed to consist solely of iron and nickel. It probably contains many other elements, but it is certainly not like any familiar solid substance because it is at a temperature of at least 4,000°C. (far above the ordinary melting-point of all known substances, and 1,000°C. above the normal boiling point of iron). It is prevented from liquefying only by the enormous pressure of 24,000 tons per square inch. This inner core is

The Lithosphere 51

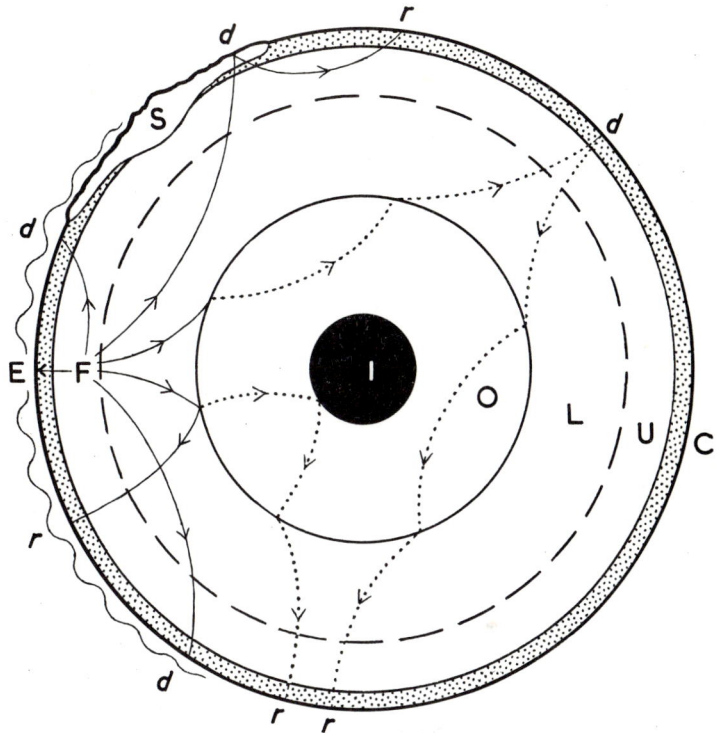

Fig. 7. Section through the earth, showing paths of earthquake waves originating at the focus F and the epicenter E. Seismographs receive direct waves at *d* but reflected waves at *r*. Continuous lines: P and S waves. Dotted lines: P waves only. Wavy lines: L waves. I, Inner core. O, Outer core. L, Lower mantle. U, Upper mantle. C, Crust (thickness exaggerated about × 15). S, Continental block.

surrounded by an outer core about 1,250 miles thick, consisting probably of similar materials but in the molten state, though the pressure is still around 18,000 tons per square inch.

The atoms in both parts of the core are so crushed by the pressure that their state cannot be normal. Some of their electrons are probably torn off and exist in the free state, so that the material has metallic properties. The inner core, though it may be rich in iron, cannot be the cause of the earth's magnetism because even pure iron is not magnetic at that temperature, but the fluid outer core is believed to circulate round the earth and to be sufficiently rich in free electrons to behave like an electric current. It is this which probably generates the earth's magnetic field, so that the earth is not a 'permanent' magnet but an electro magnet.

The outer core is surrounded by a *mantle* of solid rock about 1,800 miles thick. The junction between the outer core and the mantle is marked by an abrupt drop in the pressure to 10,000 tons per square inch, the boundary being known as the *Gutenberg* or *Oldham discontinuity* (after B. Gutenberg and R. D. Oldham). This sudden change probably means that the atoms of the material are now able to assume a more or less normal state.

The mantle is the site of the slow convection currents believed to be responsible for the continental drift. It is usually divided into a lower or deep mantle, 1,230 miles thick, and an upper mantle about 590 miles thick, a slight discontinuity having been detected between them. In the upper mantle, which contains the deepest earthquakes (720 miles down), the pressure drops steadily from about 2,600 tons per square inch (at a depth of about 620 miles) to 60 tons per square inch only 33 miles from the mean surface of the earth.

At this level there is another sudden drop in the density of the rocks. This is called the *Mohorovičić discontinuity*, after A. Mohorovičić who discovered it in 1906, and it forms the boundary between the mantle and the earth's crust. As we have seen, this consists of a shell of sima in which the continents of sial 'float' like rafts, but with their roots extending down into the mantle. In the centers of the continental areas the crust may be as much as 50 miles thick, but over the ocean floors its depth is probably only a few miles.

The Lithosphere

The 'Mohole' Project

The material of the upper mantle is believed to consist of some such rock as the dunite found in Dun Mountain, New Zealand, for this is the densest rock known and consists almost entirely of the very heavy mineral olivine, which is a silicate of magnesium and iron. Other olivine rocks occur only where they have clearly been brought up in the molten state from great depths, though the few specimens found in deep oceanic rifts have been doubtfully claimed to be actual fragments of the earth's mantle.

At the International Geophysical Congress at Ottawa, Canada, in 1960, it was proposed to sink a bore hole in the ocean floor where it was thought to be about 3 miles thick. The purpose was to obtain an indisputable specimen of the mantle. 'Mohole' was the code name for the International Upper Mantle Project, and was formed from 'Moho', a common abbreviation for the Mohorovičić discontinuity, and 'hole.' The work was intended to be done between 1962 and 1964 at a total cost of about $96 million, most of which was to be contributed by the United States.

A preliminary deep water drilling test was begun in 1961 off Guadeloupe, in the West Indies, where the sea is $2\frac{1}{2}$ miles deep. A drilling tower 100 feet high was mounted on the barge *Cuss I*, a craft 256 feet long with a beam of 46 feet. The drill, assembled in sections, had to reach down $2\frac{1}{2}$ miles before work could begin, and its weight of 67 tons was partly relieved by submerged floats. The drilling barge was kept within 200 feet of the position over the boring site by means of four independent outboard motors, which were controlled electronically to counter drift in any direction.

The diamond-tipped drill was then driven successfully through 600 feet of the ocean floor, the first 570 feet consisting of more or less consolidated ooze estimated to have taken between 10 and 20 million years to accumulate. Below this, the drill passed through a layer of gray limestone and penetrated into the basalt bedrock. To reach the mantle another 3 or 4 miles of drilling would be required, but the test was regarded as satisfactory and plans were made for the full-scale operation. However, the money ran out and the US House of Representatives refused to appropriate additional funds, so the project was finally abandoned in 1966.

A more ambitious project in the USSR was to sink five bore holes into the earth's mantle at sites on land, where depths of from 9 to 12 miles would be required. The first was begun in the Kola Peninsula, north of the White Sea, after the successful sinking of a pilot shaft, in 1967. The second followed on the Mangyshlak Peninsula in the Caspian Sea, and by 1969 both these holes had completed their 'first stage' (a depth of about 5 miles). The third site, scheduled for drilling in 1970, is near Saatly, in Azerbaijan, and the remaining two are expected to be in the Ural Mountains and the Kurile Islands, respectively. Since bore holes of this length would require five or six years for their completion, the earth's mantle is not likely to be reached before 1972.

Vulcanicity

The phenomena associated with volcanoes, geysers, fumaroles and hot springs are treated in detail in books on physical geography and geology, but the manner in which hot and molten rock can occur near the top of the mantle, and even in the crust of the earth, is a geophysical topic.

The occurrence of earthquakes is evidence that considerable stresses occur locally in the rocks of the crust and upper mantle. This is not surprising since the material of the mantle is circulating in convection currents, and moving the continental blocks along by pressure against

The Lithosphere

their deep roots. The most severe stresses would be expected to occur around the edges of the blocks, and it is a matter of fact that the great volcanoes of the world occur in chains along the continental margins, as in the famous 'ring of fire' round the Pacific Ocean.

At the root of a continental block there is great pressure against the 'rear' face, where it is being pushed along, but reduced pressure in front where the currents are pulling away from the block. Nearer the surface the opposite conditions prevail, for here the block is being forced through the more slowly moving crustal rocks. High pressure now occurs along the fore-edge, where mountains may be folded up, but there is a region of reduced pressure at the rear where the block is dragging away from the crust. The pressure is also reduced above the sinks of descending rock, and over the ascending currents where rifts are being formed. In all the regions where the pressure is reduced pockets of molten rock called *magma chambers* may form.

This may happen to both the sima and the sial, though their physical reactions are somewhat different. The melting point of basaltic rocks (sima) rises with increased pressure until, at a depth of 30 miles, it is about 1,100°C. In some places, called 'hot spots,' their temperature is already fairly near this, but the rocks will remain solid unless there is a drop in the pressure. If this happens, they will relapse into a liquid should their temperature happen to be above their melting point for the reduced pressure.

Granitic rocks (sial) behave in a similar way if they contain only about 2 per cent of water, but their melting point 30 miles down is only 900°C. and they will melt before the basalt. On the other hand, if they contain as much as 9 per cent of water their melting point falls as the pressure increases, and at 30 miles down they will melt at only 600°C. If they are in contact with rocks of higher melting-points they may dissolve these.

Thus, magma chambers are liable to form wherever pressure anomalies occur, and in some localities they may lie a very little distance below the surface. Once the rock has melted the pressure may force it into cracks and fissures in the surrounding solid rock, whence it may finally emerge above the surface as a lava flow or volcano. Again, water may percolate downward to regions where it is drawn still farther down by descending currents in the rock, and may eventually reach a magma chamber. Here it explodes into steam and a violent eruption may occur in a volcano far above. Such an explosion may cause an earthquake if there is no volcano to act as a safety-valve. However, earthquakes of this type are less common and generally less severe than those caused by the sudden slipping of the rocks along a fracture in the earth's crust.

4

The Hydrosphere

The layer of water that covers 71 per cent of the earth's surface is exceedingly thin compared with the solid shell of the mantle. Nevertheless, it averages about half the mean thickness of the earth's crust of sima and sial, and is in many large areas much deeper than the sima on which it rests. This may give an exaggerated idea of the depth of the oceans, but in fact it shows the extreme thinness of the earth's crust, which is proportionately about 1/6 the thickness of an ordinary eggshell. We have already given one illustration of the thickness of the hydrosphere (page 18), but a more vivid idea might be obtained by trying to represent the oceans on a globe of the world about 10 inches in diameter. We should need only to scrape off the varnish over the parts colored blue and then wipe them with a damp sponge. If the water wetted them evenly with a film only 1/400 of an inch thick, that would represent the mean depth of the sea. In one or two places we could make slight dents with a thumb nail about twice as deep, and these would represent the deepest trenches in the abyss. The top layer of paper beneath the varnish would do for the earth's crust as now generally defined.

By human standards these dimensions are nevertheless enormous. The seas of the world cover nearly 140 million square miles with a mean depth exceeding $2\frac{1}{4}$ miles, and a maximum depth of about $6\frac{2}{3}$ miles (36,198 feet, sounded in the Marianas Trench of the Challenger Deep by the

USSR research ship *Vityaz*, 1959). It is interesting to compare these figures with corresponding measurements for the elevation of the land above sea level. The lands of the world cover rather more than 55 million square miles, have a mean altitude of less than ½ mile, and a maximum height of about 5½ miles.

The Sea Floor

The bottoms of the great oceans are fairly flat over very large areas, and the *deeps* and *trenches* descend from them suddenly and locally, much as the continental mountain ranges rise steeply from the plains. The wide ocean bottoms are referred to as *abyssal plains*, but as they approach the margins of the continents they begin to rise and eventually slope steeply upwards. These zones are known as the *continental rise* and the *continental slope*. See Fig. 8. The continental slope usually ends about 600 feet below the surface, the sea floor then becoming flat again as the *continental shelf*, the water above which is often referred to as a *shelf sea*. The continental slope and shelf are really a submerged part of the continental block of sial, which is embedded in the sima forming the bedrock of the abyssal plains.

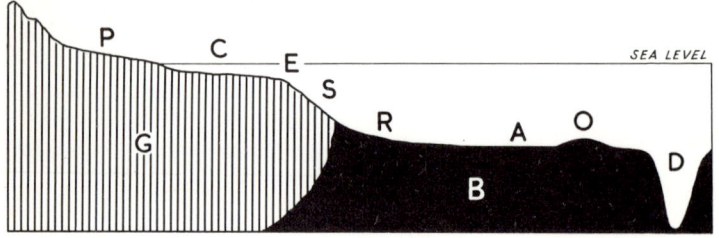

Fig. 8. Generalized section through land and sea. P, Continental platform. C, Continental shelf. E, Continental edge. S, Continental slope. R, Continental rise. A, Abyssal plain or deep-sea platform. O, Mid-ocean rise. D, Ocean deep. G, Continental block of sial (*e.g.* granite). B, Ocean floor of sima (*e.g.* basalt).

The Hydrosphere

The continental slope is said to 'rise steeply,' but this is a relative term. The average gradient is about 1/30, and everywhere it bears hills, valleys, crags, local plains, etc., much like a submarine landscape. The continental shelf, again, may be so narrow as to be scarcely recognizable, or it may be 700 miles wide as it is off northwest Europe. The British Isles stand on this shelf, of which they are merely a part that happens to project above the present sea level. Such islands are called *continental islands*, and the North Sea and English Channel are examples of shelf seas.

Other local elevations rise even from the abyssal plain, but these generally form *ridges* or *rises* (plateaux) from the tops of which a few peaks may break the surface as *oceanic islands*. The islands of Polynesia, Melanesia and Hawaii are of this kind, and so are the Azores, St. Helena and Tristan da Cunha. Well-marked peaks that do not reach the surface are called *seamounts* or *submarine mountains*, and some are active volcanoes. The highest known seamount was discovered in 1953 near the Tonga Trench, between Samoa and New Zealand. It rises 28,500 feet (5·36 miles) above the abyss, and its summit lies 1,200 feet below the surface of the ocean.

The positions of the chief ridges and trenches are shown in the map in Fig. 9, where it will be noted that they often run parallel with one another. The festoons of islands off eastern Asia are probably detached fragments of the continent, but they are also accompanied by deep trenches, possibly torn open by the same dragging movements that detached the islands. Trenches on the other sides of the continents, such as the Sunda Trench and the Peru-Chile Trench, may be situated over descending currents in the earth's mantle which are locally drawing the crustal rocks downwards (see page 44). The mid-oceanic ridges, such as the Dolphin Ridge in the North Atlantic, and rises like the South-East Pacific Plateau, are probably situated over ascending currents in the mantle, which are

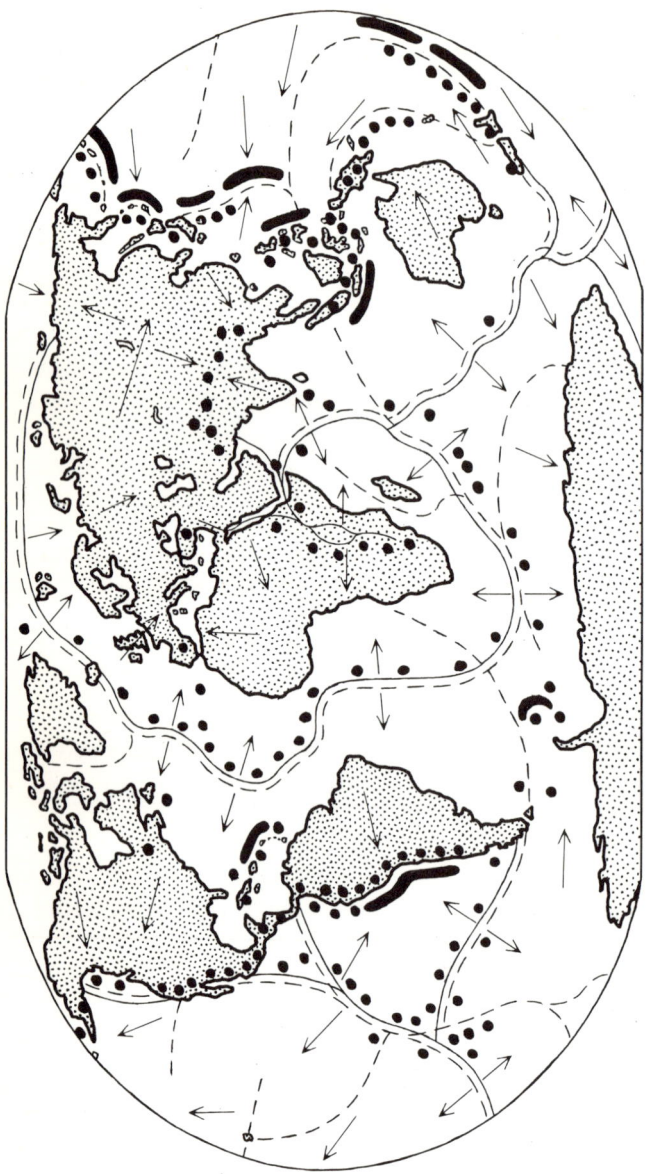

Fig. 9. Mid-ocean ridges are shown by broken lines, and the rifts which often split them by thin continuous lines. The short thick blocks indicate the great deeps or trenches, and the several dots cover the sites of numerous earthquakes. The arrows show conjectured movements of the earth's crust; note the sea floor spreading (see page 47).

The Hydrosphere

pushing the crustal rocks upwards.

Another feature of the mid-oceanic ridges is that they form a more or less continuous system of submarine mountain ranges. During the International Geophysical Year, these were discovered to be cracked down the middle for the greater part of their length. These steep-sided rifts measure from 9,000 to 15,000 feet deep (in the Atlantic), and are the site of the great majority of submarine earthquakes and volcanoes. The ascending currents that raised the ridges could also account for their cracking, but the rifts are so long that, with their branches, they virtually enclose the globe in a large network of cracks, and this has been put forward as evidence that the interior of the earth is expanding (page 47).

The nature of the deposits on the ocean floors depends largely on their distance from the land. The continental shelves, which are usually narrow on the eastern margins of the land-masses but broad on the western margins, generally support sands and clays brought down by rivers, or sands and shingles produced by marine erosion. The abyssal plains, which are beyond the range of deposits washed from the land, are covered by oozes derived largely from the *tests* (shells) of microscopic animals (Foraminifera, Radiolaria, Pteropoda) and plants (Diatoms, Coccoliths), which live chiefly in the surface-waters but sink to the bottom when they die. Large areas are also covered by the 'Red Clay,' derived from fine volcanic dust carried across the oceans by the wind.*

There are also secondary deposits on the ocean floor of a very remarkable nature. These are more or less round or bun-shaped nodules of manganese, iron, copper, nickel and cobalt ores. They occur usually in single layers on or just beneath the surface of the normal sediments. They were discovered by the famous oceano-

*For further discussion of the oceanic oozes, and of organic deposits such as coral, which often surrounds oceanic islands, the reader is referred to books on physical geography and oceanography.

graphic expedition in *HMS Challenger*, in 1872–1876, but were virtually neglected until the International Geophysical Year, 1957–1958. When sectioned, they are seen to consist of concentric layers like the sheaths of an onion. They evidently grow by accretion at an estimated rate of about one millimeter per 1,000 years.

Some of these nodules are remarkably large; one was dredged up from the eastern Pacific (in 1955) which measured about 2 feet in diameter. This may have taken half a million years to grow, and since contact with the seawater is necessary the deposition of the normal ooze must have occurred at an even slower rate. In some regions about four pounds of nodules per square foot of the sea floor are found over large areas, though one pound per square foot is more usual. The bulk of the ore is manganese dioxide, and the nodules occur at depths varying from a mile to nearly three miles. This vast store of wealth is likely to be exploited in the future, since deep-sea dredges have already been devised for its recovery. On the shallower bottoms of the continental shelves there are also masses and slabs of the mineral phosphorite (calcium phosphate), a valuable fertilizer, and it will probably not be long before the commercial harvesting of these is undertaken.

Conditions Under the Sea

Seawater, of which the average composition is given in Table IV, is not as transparent as fresh water. In bulk it appears a cloudy green instead of a clear blue, but this is partly owing to suspended particles of solids and microscopic plants and animals. The open sea, as seen from the shore or the deck of a ship, does not necessarily appear green because what is usually seen is sky-light reflected from the surface. Sunlight scattered by particles suspended in the water, which absorb the red rays, may make the sea appear deep blue or even violet, while shadows cast

The Hydrosphere

Table IV
AVERAGE COMPOSITION OF SEAWATER IN PARTS PER 1,000 BY WEIGHT

Sodium chloride (common salt)	27·213
Magnesium chloride	3·807
Magnesium sulphate (Epsom salts)	1·658
Calcium sulphate (gypsum)	1·260
Potassium sulphate	0·863
Calcium carbonate (limestone)	0·123
Magnesium bromide	0·076

Several other elements, including iron, sulphur, iodine, copper and gold, occur in traces.

by lines of waves may make the sea appear black from a distance.

Beneath the surface, the sea appears green only for the short distance to which the light penetrates without alteration. Below the first few feet absorption of the red rays leaves the sea a bluish-green, and farther down both green and blue rays are absorbed, and so little light is left that the water appears dark gray or almost black. Lower still it is too dark to support plant life, though animals continue to flourish right down to the abyss. The level at which complete darkness supervenes varies with the intensity of the sunlight or daylight above, but the mean limit is about 600 feet, or the level of the continental shelf.

The pressure, too, grows rapidly as the weight of the water above increases. The surface is at atmospheric pressure, or about 15 pounds per square inch, but this is doubled 33 feet down, trebled at 66 feet, and so on, one 'atmosphere' being added for every 33 feet. This means that on the continental shelf the pressure is about 300 pounds per square inch, on the abyssal plain it is 3 tons, and in the deepest trenches 7 tons. These values were calculated long before there was experimental evidence of them, but in the 19th century it was found that an empty sealed bottle let down into the deeps is shattered by the pressure, and a sealed can is squashed flat. The fish inhabiting these regions survive because every part of them contains water at the same high pressure. When hauled to the surface, their bodies swell up and sometimes burst from the pressure of the water locked in their tissues and organs.

The temperature of the sea varies with the season, the latitude and depth. The upper layers warm up during the summer, but water is such a poor conductor of heat that the effect does not reach its maximum until August (in the northern hemisphere). The warmth penetrates to about 200 feet deep by October or November, but below 300 feet the lag is so great that the seasons here are reversed, the winter water being warmer than the summer. However, these effects are greatly modified by currents, and below 600 feet there are no seasonal changes at all. The sea becomes steadily colder until it stands permanently at a few degrees above freezing (about 4°C) on the abyssal plains. The highest recorded summer temperature on the surface is 96°F (almost blood heat) in the Persian Gulf. The lowest possible temperature is 28°F, which is the freezing-point of seawater and occurs in the neighborhood of polar ice.

Ocean Currents

The difference in temperature between the polar and tropical waters is the major cause of ocean currents, for the cold heavy polar water sinks down below the warm water, which flows towards the poles to take its place. Thus a system of convection currents is established. On the surface (in middle latitudes) the poleward flow of tropical water lies beneath the belt of permanent winds known as the Westerly Variables. Like the winds, the water is deflected by the earth's rotation (see page 89) and on the western margin of the North Atlantic, off the coast of the United States, both wind and water are

The Hydrosphere

moving in the same direction. The wind blows faster and accelerates the speed of the surface water, about 3 knots being attained by the Gulf Stream. On the eastern side of the ocean, off the coast of North Africa, the Trade Winds similarly blow the surface water towards the southwest, so that the whole of the North Atlantic surface-water is kept moving in a vast circle by the Westerlies and the Trade Winds, which turn it like two men on opposite sides of a capstan. It is 'kept moving' by the winds, but it is the earth's rotation that forces it to go in a circle. These effects are clearly displayed in the map of the ocean currents given in nearly all atlases.

Superficial currents caused by the winds in this manner are called *drifts*, and they are of great importance both to navigators and to the countries whose shores they wash, for they influence the climate of all the adjacent land. The comparative mildness of the British climate, for example, is due to the North Atlantic Drift, which is continually supplied with warm water from the Gulf Stream. The circular current sweeping clockwise round the North Atlantic encloses a large area of nearly still water called the Sargasso Sea, in which seaweeds flourish and flotsam collects, much as floating tea leaves gather in the center of a cup of tea that has been stirred. There is a similar tendency in other oceans but with much smaller effects.

The circulation below the surface is more complex, and is influenced by variations in the *salinity* or saltiness of the sea. The mean salinity is about 3·5 per cent by weight of dissolved salts (mostly sodium chloride or common salt), but in hot dry latitudes where there is excessive evaporation, and especially in land-locked areas like the Mediterranean Sea, the solution becomes more concentrated and the sea-water more dense. This water sinks to the bottom, and in the Mediterranean pours out through the Strait of Gibraltar as an undercurrent and spills over the Atlantic sea floor. Here, at a depth of one or two miles, it meets the much colder polar water traveling southwards from the Arctic, mixes with it and warms it. This bottom water continues to travel southwards until it meets the cold polar water from the Antarctic and, being warmer, rises over it and 'upwells' to the surface, where it joins the surface circulation.

When the sea freezes in the polar regions the ice formed is of fresh water, its salt being left behind to increase the salinity of the sea. This contributes to the sinking of the polar water. When the ice melts in the summer season, a great deal of fresh water is released at the surface so that the deeper and saltier water is again pushed downwards. In other regions the salinity of the sea is greatly reduced by heavy rainfall, such as occurs in the equatorial rain belt, or by the outflow of great rivers like the Amazon. Becalmed sailors are said to have died of thirst in the Atlantic Ocean 300 miles off the mouth of the Amazon, not realizing that the water even at this distance may be still fresh enough to drink.

In all these circumstances the lighter fresh water tends to ride over the salt water, and the salt water to sink below the fresh, adding complexities to the circulation that geophysicists are still engaged in unraveling.

Another type of modification is caused by tidal currents, which will be explained in connection with the tides, and there are very likely unsuspected features of a major character awaiting discovery. One such turned up in 1951, when a strong easterly current was found flowing beneath the surface of the Pacific Ocean along the equator. Named the Cromwell Current after Townsend Cromwell who first studied it, it travels at nearly 3 knots at a depth of 300 feet, passing about 40 million tons of water per second, and is the second largest known current in the Pacific Ocean.

The Hydrosphere

The Tides

The tides are periodic changes in the level of the sea caused by the gravitational attraction of the sun and moon, but the moon is so much nearer the earth that, in spite of its small size, it exerts about $2\frac{1}{4}$ times the sun's tide-raising force. The manner in which the tides are raised is by no means simple, for the moon not only raises the water on the side of the earth facing it, giving one daily high tide, but it also raises the water on the opposite side of the earth and thus provides a second daily high tide. See Fig. 10. Since the total quantity of water remains unchanged, these tidal 'bulges' are separated by regions of low tide, where the level of the water is below normal.

The explanation commonly given is that the water on the side of the earth facing the moon is nearly 4,000 miles nearer the moon than is the earth's center, and is therefore attracted more strongly. Being a liquid, it yields to this attraction and becomes heaped up on the side of the earth facing the moon, and places in this hemisphere then have a high tide. So far so good, but the explanation of

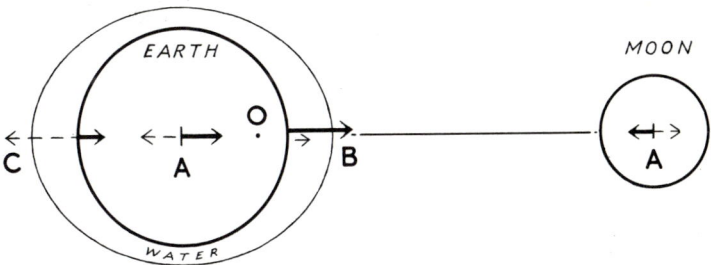

Fig. 10. Ideal explanation of the tides. *Thick arrows:* gravitational attraction. *Broken arrows:* centrifugal force owing to the revolution of the earth and the moon about the common centre O. A, Attraction and centrifugal force exactly balanced at centres of the earth and moon. B, Attraction strongest on side nearest the moon, supplemented by weak centrifugal force; tidal bulge raised. C, Weak attraction on side away from the moon overcome by strong centrifugal force, which raises the second tidal bulge.

the high tide that simultaneously occurs on the opposite side of the earth is not satisfactory. It states that the water there is 4,000 miles farther away from the moon than is the earth's center, and is therefore attracted *less* strongly. Since the earth is thus being pulled more strongly than the water, this gets left behind and so forms the high tide on the far side. It is unsatisfactory because the earth does not *in fact* move towards the moon at all, and so the water cannot really be said to be 'left behind.' Both tides are better explained in a different way.

It is usually said that the moon revolves about the earth once every month, but it is more accurate to say that the moon and earth revolve about *each other* once every month. Since the earth is about 80 times as heavy as the moon it describes only a very small circle, while the moon makes a wide sweep round a large orbit. The common center of both orbits actually lies inside the earth, about 1,000 miles below the surface, so that the earth's motion has rather the character of a wobble.

We may compare the system to a very fat man swinging a small boy round and round by his arms, but we must imagine that the boy is holding onto the front of the man's coat, and that the man is not using *his* arms at all. The boy flies round and round through the air in a circle, pulling the man's coat out to make one tidal 'bulge.' Since the man has to lean backward slightly to keep his balance, he also wobbles round in a small circle. Now, suppose he is wearing a coat with 'tails.' If he goes round fast enough his coattails will be flung out behind by centrifugal force, and it is in a similar way that the water on the side of the earth away from the moon is flung out to produce the second tidal bulge.

The tides raised by the sun can be explained more clearly in yet another way. The earth goes round the sun at about 18 miles per second. If it traveled more slowly it would fall into the sun; if it traveled more quickly it would fly away into space. A nice balance is preserved,

The Hydrosphere

the general rule being that a greater speed is required to keep bodies in orbit when they are nearer the sun, but a slower speed when they are farther away.

Now, the earth moves round the sun carrying the hydrosphere with it, but on one side the water is 4,000 miles nearer the sun than is the center of the earth, so to avoid falling into the sun it ought to be traveling at rather more than 18 miles per second. This, of course, is impossible, so it tends to fall sunwards and produces a high tide on that side of the earth. The water on the opposite side is 4,000 miles farther away from the sun than is the center of the earth, so to avoid flying away into space it ought to travel at rather less than 18 miles per second. This also is impossible, so it does tend to fly away and produces a high tide on the far side of the earth.

We now have to consider how the sun and moon work together. At new moon and full moon the earth, moon and sun are all in a straight line, so that the tides produced by the sun are simply added to those produced by the moon. This results in extra high tides called *spring tides*. At the times of 'half' moon (or first and last quarter) the sun and moon are pulling on the earth at right-angles, and the sun then tries to raise its more feeble high tides where the moon is producing low ones. The moon wins, but the sun does have some effect and the tides are everywhere less strongly marked. These are called *neap tides*.

This sounds very simple, but there are complications even here. The solar tides have an interval of exactly 12 hours, but the lunar tides come at intervals of 12 hours and 26 minutes owing to the motion of the moon round the earth, so that they periodically get out of step. However, the moon always dominates the situation. The sun actually scores about 22 feeble tides to the moon's 21 strong ones, but we do not propose to bother with the solar tides any further.

The tidal bulges remain in line with the moon, which

moves so slowly round its orbit that we may think of it as stationary. But the earth spins round once every 24 hours, carrying the hydrosphere with it, so that we should expect a sort of 'wave of bulging' to sweep across the oceans as they pass beneath the moon. The tide certainly rises and falls like a wave, and not like a rush of water up a hill and down the other side, and if the earth were entirely covered with water each bulge would cross the oceans like a wave crossing a pond.* The proper name for the bulge is therefore a 'tidal wave,'† but though the distance between the two waves is about 12,000 miles (on the equator), their rise and fall in mid-ocean is only one or two feet.

The earth, however, is not entirely covered with water and, except in the Southern Ocean, the tidal waves are prevented from sweeping round the world by the continents. The Pacific tides come up against Asia and Australia, the Indian Ocean tides meet Africa, and the Atlantic tides are interrupted by America. It looks as if a new tidal wave must be formed to cross each great ocean, but this is not quite what happens.

The tides have been rising and falling in the ocean basins for so long that they have become what are called *standing waves*. That is to say, they flop sedately up and down, the middle rising when the sides fall and then falling when the sides rise, without bothering to move along at all. See Fig. 11. They do this in separate areas of the oceans where the natural period of the water's oscillation (that is, 'flopping') is about 12 hours and 26 minutes. The pull of the moon as it passes merely keeps them oscillating, one after another, in their natural periods. Each system is modified by the rotation of the earth, and where we might expect a continuous crest of high water we find it

*The *water* itself does not cross the pond, for a floating cork is not carried along but merely bobs up and down as the wave passes.
†The expression 'tidal wave' is often misused for large waves caused by earthquakes, which sometimes sweep across the land and do spectacular damage. The proper name for these waves is *tsunami*. They have nothing whatever to do with the tides.

The Hydrosphere

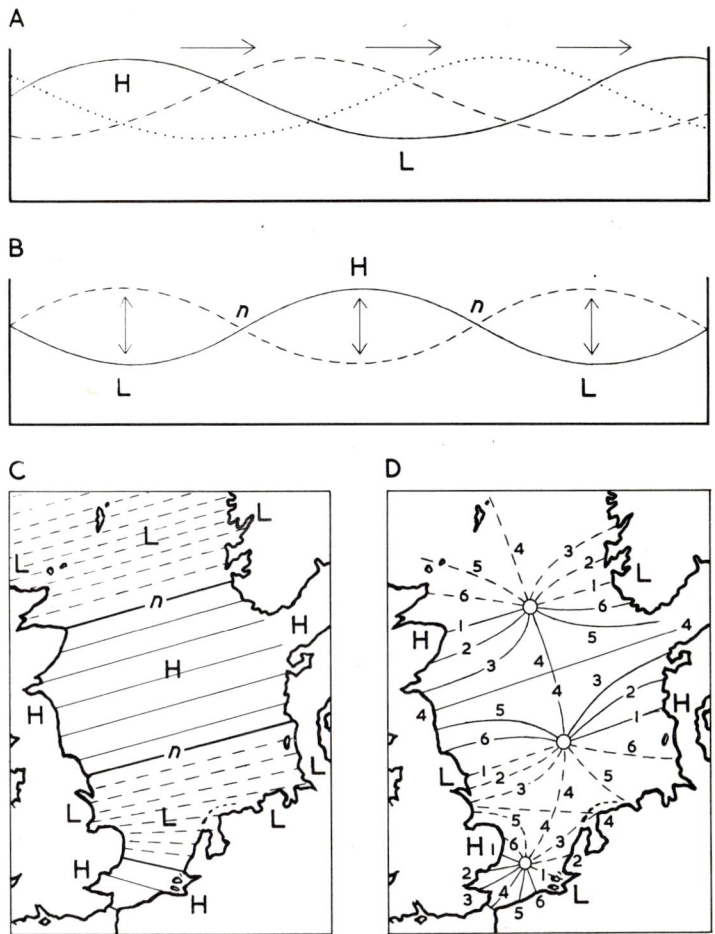

Fig. 11. A, Progressive waves travel along, as shown by the arrows, and all the water is subject to movement. B, Standing or oscillating waves flop up and down, and the water at the nodes *n* remains at the same level. C, Oscillating areas in the North Sea, disregarding effects of the earth's rotation. D, The same modified by the earth's rotation, the positions of six successive tides being numbered in order. Note that at tide No. 4 the situation approximates to that shown at C. High tides: H, continuous lines. Low tides: L, broken lines.

high at one end but low at the other. The 'flopping' is turned inside-out, so to speak, about halfway along, at an *amphidromic* or neutral point where no tides are apparent. The oscillations then cause the tides to appear to rotate about the amphidromic point in a counter-clockwise direction in the northern hemisphere, but a clockwise direction in the southern.

The large oceanic systems set others oscillating in gulfs and partly enclosed seas, and these may 'pass it on' to yet others, often with highly complicated results. An outline of the systems generated in the North Sea by the Atlantic tides is indicated in Fig. 11. In spite of its difficulties, the standing-wave theory explains the observed tide tables much more satisfactorily than the older theory, which held that the tidal wave in the Southern Ocean turned northwards when it met South America, for example, and simply ran up the Atlantic Ocean. Waves do not turn in that manner, and though currents may, there is no such current in the Atlantic. The tides we actually receive are thus only secondary effects of the basic causes illustrated in Fig. 10, which represents the ideal for a planet covered by water.

Tidal Currents

When a tide rises at the mouth of a large gulf or inlet in the coast, other phenomena may occur. In one, the standing wave generates a *progressive* wave which travels up the gulf, and if this is both wide and shallow the wave loses energy by friction with the bottom, and the height of the tide steadily decreases. On the other hand, if the gulf rapidly narrows and the water is deep enough, the tidal wave is constricted more and more as it advances and its height increases. In these circumstances the raised water may flow bodily forward as a tidal current. This happens, for example, in the English Channel, up which a tidal current flows at about 2 knots and raises

The Hydrosphere

tides along its shores which measure 16 feet or more between high and low water.

The tidal current flowing into the narrow funnel of the Bristol Channel and Severn Estuary moves at 10 knots and produces tides rising 40 feet. The excess of water finally advances up the Severn River with a wall-like front of tumbling waves several feet high, known as the Severn 'bore.' An advancing tidal current or 'flow' is succeeded by a retreating current or 'ebb' in the opposite direction as the period of low tide approaches. In all cases the tidal currents are affected by the earth's rotation, tending to curve in a clockwise direction in the northern hemisphere, and in a counterclockwise direction in the southern. This causes the tides on the French side of the English Channel to be rather higher than those on the English side.

The highest tides in the world are said to be those in the Bay of Fundy, Nova Scotia, where the difference between high and low water is commonly 60 feet. But these tides are caused in a different manner, the rise and fall of the oceanic standing wave generating a secondary standing wave in the inlet. The volume of water in the Bay of Fundy has a natural period of oscillation which makes it susceptible to resonance, and it is set in motion by a process similar to that of 'rocking the boat.' The same kind of phenomenon occurs in Long Island Sound, off the Connecticut shore.

Deep-sea Sounding

Sailors have measured the depth of the water beneath their ships by means of 'lead lines' or 'plummets' from time immemorial, but this method could not be used for the ocean deeps until steel cables were invented. The mere weight of a hempen line a few miles long would break it at the top, but steel cables can be made to sustain themselves, and, though they stretch, allowance can be made for this. An enormous number of plummet soundings of the ocean deeps had been made by *HMS Challenger* and other research ships by the end of the 19th century, and the Kelvin deep-sea sounder, invented about 1870, enabled specimens of the ooze on the sea floor to be brought up in exchange for the weights let down.

Since then, the ocean floors have been mapped in detail over vast areas by *sonar* or 'echo sounding.' In this, a bell is gonged on the bottom of a ship, or a small charge of explosive is detonated, and the echo from the sea floor is picked up by a microphone. The time interval between the signal and the echo provides a measure of the depth, though the calculation is by no means simple. The average speed of sound in seawater near the surface is about 4,742·5 feet per second, but it increases by 95 feet per second per mile of depth, and by about 43 feet per second for every one per cent increase in salinity.

Sonar has by no means abolished plummet sounding for special purposes. In the 1950's a type of sounder devised by E. Bullard and A. E. Maxwell of the Scripps Institute of Oceanography, California, was used for collecting samples from the ocean floor in order to determine their heat conductivity, and an apparatus equipped with a sensitive probe was driven into the ocean bed to ascertain the heat-flow values described on page 28. Other types of sounder have been used to determine the penetration of cosmic rays and solar radiation, and others again to detect the flow of submarine currents at all depths.

An ingenious new type of 'free' sounder for detecting light currents on the ocean floor was successfully tried in 1966. Three aluminum balls, measuring 22 inches in diameter and packed with recording instruments, were dropped into the Pacific Ocean off California and came to rest $2\frac{1}{2}$ miles down on the bottom. In this high-pressure environment they are comparatively light and easily rolled along; they have, in fact, been moving at about

The Hydrosphere

120 feet per hour. This is about 80 times slower than many surface currents, and 176 times slower than a man's normal walking pace, yet this gradual bottom movement has been shown to be significant in the slowing down of the earth's rotation by friction.*

These 'permanent sea-floor geophysical recording stations' do much more than detect bottom currents. They also contain instruments to measure the temperature and pressure of the water, the propagation of faint sounds in the abyss, the long-period vibrations of distant earthquakes, and even the very small tremors caused by sea waves in storms far above. They send up a continuous stream of data and their designer, geophysicist Gary Latham, considers that a sufficient number could be used to give warning of approaching typhoons and *tsunami*.

Exploration Downward

The world under the sea has always aroused man's curiosity, and it was doubtless inspected by divers in prehistoric times. Naked divers cannot go down very far, though pearl and sponge divers have been known to descend 120 feet. (The record depth of 200 feet was attained by Stotto Georghios, of Greece, in the Adriatic Sea in 1913). Divers have brought up innumerable interesting things from the sea floor. Many more have been recovered by dredges, trawls and deep-sea sounders, but detailed examination of the bottom had to wait until men found means of staying underwater much longer than they could hold their breath.

The first device with this purpose was the diving bell, described by Aristotle about 330 BC and said to have

---

*The earth is said to lose about 2 seconds every 100,000 years by tidal friction, but its rotation is known to be irregular. During some periods the loss has been much greater, although the rotation seems to have remained steady throughout the Mesozoic era. Causes of the fluctuations may include earthquakes and possibly the expansion of the globe (see page 48).

been used by Alexander the Great. The diving bell was to have a very long history, and finally to evolve into the submarine laboratory, but meanwhile a succession of diving dresses equipped with various means of supplying air for breathing made their appearance.

The first practical helmeted diving dress was invented by Augustus Siebe in 1819, and was improved by him in 1857. Air was pumped down to the helmet through tubes, allowing the diver to do useful work at a depth of 200 feet. Such diving dresses are still used for inspecting and repairing damage to ships below waterline and for salvage work on sunken wrecks. On the other hand, diving bells holding several men are preferred for civil engineering work down to about 35 feet.

The main difficulty to be overcome in deep diving is the effect of the pressure on respiration. Under pressure the blood dissolves some of the nitrogen in the air, and when the diver ascends the gas is evolved in small bubbles so that his blood becomes froth. This causes a paralyzing and painful cramplike condition known as the *bends*, which may be fatal. Until the invention of the submersible decompression chamber (by Robert Davis) divers had to come up in easy stages to allow the slow release of the gas. After an hour's submersion at a depth of 200 feet, for example, the ascent had to be dragged out through four hours.

The submersible decompression chamber is a kind of 'lift' which the diver enters while still on the sea bottom. He is then hauled rapidly to the surface, where he can be decompressed in comparative comfort, either in the chamber or in a room connected with it. Here, he can take off his helmet and even sleep if necessary. The short dives that have been made to 500 feet would have been physically impossible without such a chamber, for they demand a decompression period of three days.

The danger of accident to a diver's air pipes long ago led to the alternative practice of strapping steel bottles of

The Hydrosphere

compressed air to the diver's back. In recent years it was realized that if a mixture of oxygen and an inert gas like helium, which is only two-fifths as soluble as nitrogen, were used instead of air, the period of decompression and the risk of 'bends' could be greatly reduced. The diver would also avoid 'nitrogen narcosis,' or loss of consciousness caused by nitrogen in the blood.

Helium has its own drawbacks, but they are of a much less serious nature. For example, it is much lighter than nitrogen and has the effect of raising the pitch of the voice, which may become so high as to be unintelligible without the use of a microphone and frequency reducer ('unscrambler'). Helium also accelerates the loss of body heat and so reduces the permissible period of immersion in cold water, but this has been overcome by a simple hot-water heating system in some types of dress. It has been found that the use of helium instead of nitrogen enables a diver who spends about 30 minutes at a depth of 380 feet to be decompressed in about 3 hours.

Of course, it is not very economical to spend 3 hours out of every $3\frac{1}{2}$ being decompressed, but it has been discovered that if a diver remains below until his blood has taken up all the gas it can absorb at that pressure, a prolonged stay does not increase the time required for decompression. Means have therefore been devised to enable a diver to remain very much longer beneath the sea, and even to live there for several days.

This general idea is by no means new, and as long ago as 1648 Bishop Wilkins considered the possibility of fixed submarine residences of such a size that whole communities could live permanently in them. In 1899 L. de Rigaud patented a complete underwater house of six stories for residential occupation, but it was not until the 1960's that submarine residences became practicable.

The pioneer experiments were performed by E. Link from 1956 to 1962, and in 1962 and 1963 J. Y. Cousteau resided several times on the sea floor for periods of up to a month. Then, in 1964, Link came up with his 'SPID' (Submersible Portable Inflatable Dwelling). This was a sort of rubber ballon with a wide open neck underneath, and was inflated with a mixture of 4% oxygen, 91% helium and 5% nitrogen,* much as a water-spider's nest is filled with air. In this, J. Lindbergh and R. Sténuit camped for 49 hours at a depth of 432 feet. In the same year G. Bond, of the United States Navy, remained 11 days at a depth of 192 feet off Bermuda in *Sealab I*, a cylindrical steel capsule 40 feet long. In 1964 Cousteau established an underwater 'village' on the bottom of the Red Sea. Designated *Conshelf 2*, this consisted of 4 cylindrical houses assembled in the form of a cross at a depth of 35 feet, with a 'garage' to hold a diving vehicle shaped like a 'flying saucer,' and a sub station at a depth of 85 feet with living quarters for two men. In 1965, Cousteau lived for 3 weeks at a depth of 330 feet in the Mediterranean Sea, using *Conshelf 3*, a steel sphere containing two stories, and in the same year the Russians occupied a laboratory at a depth of 45 feet in the Black Sea, preparatory to establishing a station at a depth of 975 feet.

Another 1965 project was the American *Sealab II*, a steel cylinder 57 feet long divided into a laboratory and living quarters for 11 men, and carrying enough breathing gases and food for a residence of 6 weeks. As in all other such structures, provision was made for personnel to don breathing equipment and swimming flippers, and emerge through hatches and shark-trap gates to make exploratory excursions, or to engage in salvage work. Here, scientists studied the effects of submersion on human physiology, and the currents and geology of the sea-bottom off California. They also built a large fish-cage to enable tests to be made of the internal gases in fishes at a depth of 210 feet. Another experiment was the use of dolphins to

*The percentage of oxygen is reduced because less is required under pressure and an excess has toxic effects.

The Hydrosphere

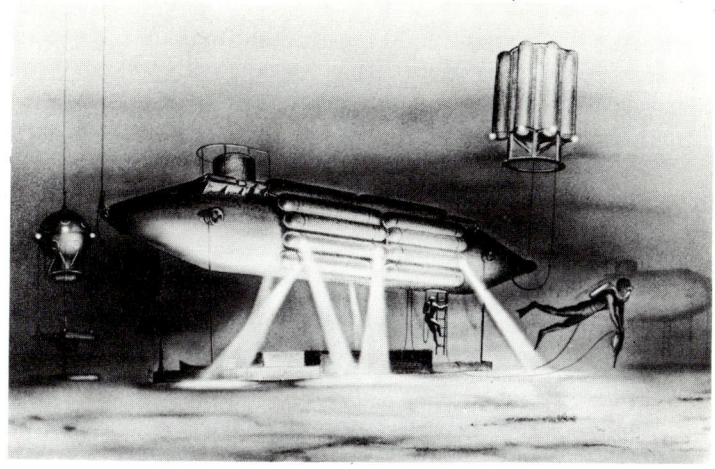

Fig. 12. An artist's impression of *Bacchus*, an underwater residential laboratory for scientific and technological research or a workshop and base for salvage operations. (*British Aircraft Corp.*)

carry messages to and from the surface. The remarkable intelligence of dolphins may enable them to be trained to carry letters and packages from one submarine station to another, but the 'dolphin post' is still in the future! Fig. 12 shows a typical underwater dwelling.

Several similar projects have been undertaken since, including *Sealab III*, intended to be occupied in 1969 by relays of 'aquanauts' in five teams of nine each. Representing four countries, the teams were expected to spend 12 days each at a depth of 600 feet, with occasional excursions down to 1,000 feet. Other 1969 projects include the underwater laboratory *Tektite I*, and the residential workshop *Bacchus* (British Aircraft Corporation Commercial Habitat Under the Sea).

The names *Sealab* and *Conshelf* indicate the long-term object of these projects, which is to establish permanent laboratories and workshops on the continental shelf, and generally 'open up' this vast unexplored territory. More ambitious projects include large underwater industrial plants for the systematic harvesting and partial processing of the mineral wealth of the sea bed. Prospecting can already be done with several types of mobile machines rather like small submarines, but fitted with external grabs and other tools. Although these were developed chiefly for servicing underwater oil wells, they are adaptable to oceanographic survey work.

Into the Depths

The devices so far described enable men to live and work under the pressure prevailing on the continental shelf. See Fig. 13. At greater depths respiration and movement become increasingly difficult, though in 1962 H. Keller and P. Small left a diving bell for a brief spell at a depth of 1,000 feet. The pressure at this depth is nearly 500 pounds per square inch, and this record cannot be far from the absolute limit of human endurance.

The only way to overcome this limitation is to provide pressure-proof containers in which a man can live at normal atmospheric pressure, and to provide windows for observation and perhaps external tools which can be manipulated from inside. The first success in this direction was the German deep-sea diving dress, constructed of steel and aluminum, which enables a diver to do salvage work at a depth of 750 feet with a pressure outside the dress of about 300 pounds per square inch. The provision of jointed arm and leg tubes allowed limited freedom of movement, and claspers at the ends of the arms could be worked by the hands inside.

At pressures of more than 3 hundredweight per square inch a jointed dress is impracticable and a completely sealed spherical capsule is necessary. Such a capsule forms the habitable part of the oceanographic research vessel *Alvin*, which is a miniature egg-shaped submarine capable of exploration down to 6,000 feet, where the pressure is

The Hydrosphere

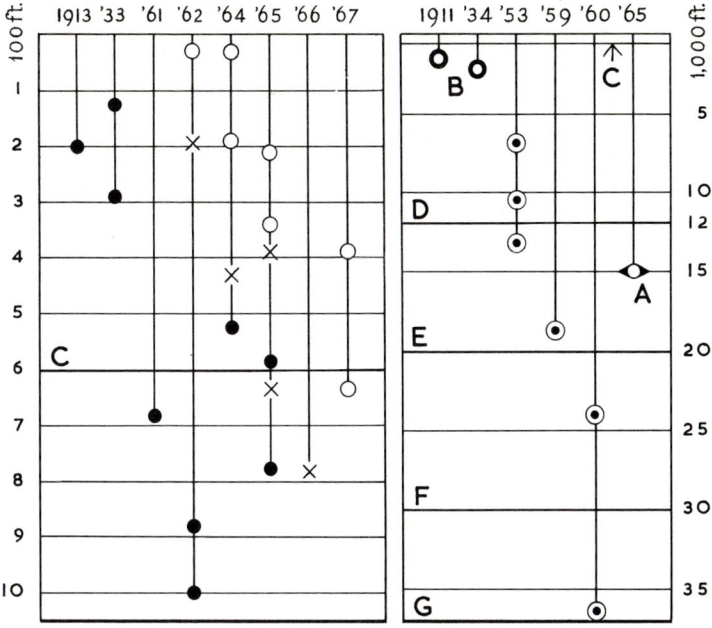

Fig. 13. Some outstanding oceanic descents. C, Continental shelf. *Left:* The first 1,000 feet. Dots: Dives of less than one day. Crosses: Spells of between one day and one week. Circles: Residences of more than one week. *Right:* Down to the bottom. B, bathyspheres. Dotted circles: bathyscaphes. A, *Aluminaut*. D, mean depth of sea. E, more than 80 per cent of the ocean floor lies above this level. F, average deeps. G, Marianas Trench.

$1\frac{1}{4}$ tons per square inch. *Alvin* was commissioned in 1964 purely for research purposes, but in 1965 the 51-foot *Aluminaut*, able to operate at 15,000 feet, was completed. This vessel is equipped with mechanical arms 9 feet long, and was designed chiefly for mineral prospecting and salvage work.

The pioneers in oceanographic research at depths exceeding 1,000 feet were C. W. Beebe and O. Barton. In 1930, they were lowered on a steel cable in their *Bathysphere* to a depth exceeding a quarter of a mile

(1,426 feet) in the Caribbean Sea. The *Bathysphere*, a steel ball able to withstand a pressure of at least 5 tons per square inch, was fitted with two circular windows of quartz 3 inches thick and a searchlight. On its fourth trip, in 1934, it descended about two-thirds of a mile (3,028 feet), but this was still a long way from the bottom.

The risks involved in descending much farther on a cable attached to a vessel rising and falling on the surface were too great, and after World War II a new method was tried. The steel ball was attached to the underside of a large tank filled with gasolene, which would act like the gas-filled envelope of an airship. This tank was of thin metal and open to the ocean at points on its bottom, so that the pressure would be exerted on the inside as well as outside, enabling it to survive like the deep-sea fishes. Gasolene is lighter than water, so that the *bathyscaphe*, as it was called, would rise to the surface unless

Fig. 14. The American bathyscaphe *Trieste*, designed by J. Piccard. (*U.S.I.S.*)

The Hydrosphere

weighted with ballast. This was supplied in the form of iron ingots held to the bottom by magnets which could be switched off to release it. The bathyscaphe could thus be maneuvered up and down exactly like a ballon, while horizontal motion was achieved by means of a propeller. See Fig. 14.

In 1953, G. S. Houet and P. H. Willm descended $1\frac{1}{4}$ miles in a bathyscaphe off Toulon, in France, and in the same year Auguste and Jacques Piccard attained a depth of nearly 2 miles (10,335 feet) in the Tyrrhenian Sea in the bathyscaphe *Trieste*. In the following years the abyssal plain was reached, and even the ocean deeps were visited. The record – still standing in 1968 – was achieved in 1960, when J. Piccard and D. Walsh in a modified *Trieste*, entered the Marianas Trench to a depth of $6\frac{3}{4}$ miles (35,802 feet), a feat which was repeated by the *Archimède* in 1961, see Fig. 13. Thus has man invaded the hydrosphere, though the scientific work accomplished has been mainly concerned with the biology of the deeps as observed through windows. This, however, is all a part of oceanography, which is itself a branch of geophysics.

5

The Atmosphere and Beyond

The atmosphere is sometimes described as a great ocean of air, on the bottom of which we are obliged to spend most of our lives like creatures living on the bottom of the sea. It is, however, unlike the ocean of water in many important respects. To begin with, air is a mixture of gases, and gases are readily compressible while water is not. The water in the ocean deeps, though supporting several tons per square inch, is not very much denser than the water on the surface, but the density of the atmosphere decreases rapidly with altitude, becoming thinner and thinner until it eventually fades away into nothing. Thus, the atmosphere has no kind of 'surface' to match the surface of the sea.

Again, the composition of the water in the oceans is more or less the same at all depths, but that of the atmosphere changes at various ill-defined levels so that it can be divided into several concentric spheres with different characteristics. And yet again, the circulation of the ocean proceeds throughout all its levels, and especially near the top, whereas the main circulation of the atmosphere is confined to a shallow layer at the very bottom. Since we live on the literal rock bottom of this layer, it is natural that our study of the atmosphere should begin here.

The Atmosphere and Beyond

The Troposphere

This bottom layer is called the *troposphere* (from the Greek for 'sphere of change'), for it is the region of the permanent winds and the weather. Its thickness varies from about 7 miles at the poles to 10 miles over the equator, but there are at least 400 miles of detectable air above it. It consists chiefly of the gases oxygen and nitrogen, and its composition is given in Table V.

The air at any level has to support the weight of all the air above it, and therefore exerts considerable pressure. At the surface of the earth the mean atmospheric pressure is about 14·7 pounds per square inch, and a cubic foot of air weighs about $1\frac{1}{4}$ ounces. The weight of the air varies with its content of water vapor, the presence of which lightens it. It is also lightened by expansion as its temperature rises, and this is the main cause of the circulation of the atmosphere.

Table V

COMPOSITION OF THE LOWER ATMOSPHERE IN PARTS PER 1,000 BY VOLUME

Nitrogen	781
Oxygen	210
Argon	9·3
Carbon dioxide	0·3
Neon	0·02
Helium	0·005
Methane (marsh gas)	0·002
Krypton	0·001
Nitrous oxide (laughing gas)	0·0005
Hydrogen	0·0005
Ozone	0·0004
Xenon	0·0001

Water vapor may be present in any quantity up to about 50 parts per 1,000, but seldom exceeds more than a third of this in temperate latitudes. In the neighborhood of towns there are also traces of sulphur dioxide, sulphuretted hydrogen and ammonia, while thunderstorms produce small quantities of the oxides of nitrogen and a local increase of ozone.

Gases are poor conductors of heat, and a large volume cannot be warmed (in any effective time) except by convection currents. Gas actually in contact with a hot surface becomes hot and expands. If the hot surface is *underneath* a volume of gas the heated layer immediately rises. The surrounding gas then flows in to take its place and becomes heated in its turn, and this continues until all the gas has made contact with the hot surface. The sun's heat rays do not heat the atmosphere from above, but penetrate to the earth's surface where they heat the ground. The hot ground then heats the air in contact with it and starts up convection currents in the manner described. See Fig. 15.

The hottest part of the earth is known as the *heat equator*. This is the belt of land and sea lying more or less directly beneath the sun, which stands overhead on the geographical equator only at the equinoxes (March 21 and September 21). In June the heat equator follows the sun northward towards the Tropic of Cancer, and in December it moves south towards the Tropic of Capricorn. It thus forms a belt round the earth which shifts north and south with the seasons, but the distance it moves is affected by the positions of the great land masses so that it has a somewhat wavy form.

Wherever the heat equator happens to be, the air above it is strongly heated and because it expands it rises like a hot-air balloon. Since it is expanded the atmospheric pressure is reduced, and since it is moving upwards instead of sideways there are practically no winds blowing. Thus, there is a permanent belt of low-pressure calms round the earth approximately on the equator, or a little to the north or south of it. These calms were called the *Doldrums* in the days of sailing ships and the term is still used. Since the Doldrums lie mainly over the oceans the ascending air is saturated with water vapor. As it rises it begins to cool and its moisture condenses and falls as rain. This explains the *equatorial rain belt* which

The Atmosphere and Beyond

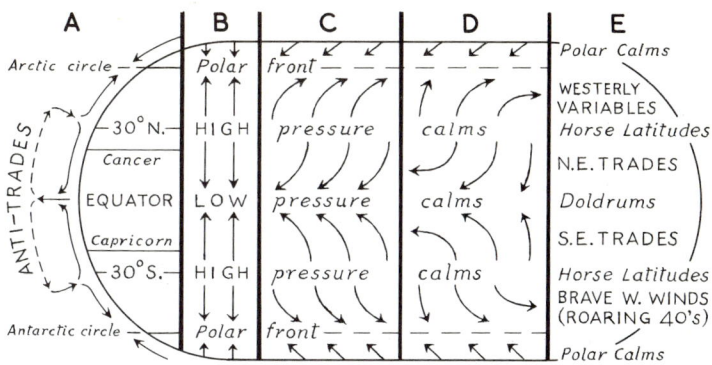

Fig. 15. The planetary wind system. A, Circulation of the atmosphere. B, Surface winds ideally conceived. C, Modifications owing to the rotation of the earth. D, Further modifications over an ocean caused by land-masses to the east and west of it. E, Traditional names. To picture the circulation of the Trade Winds and Anti-trades in correct proportion, imagine either of the arrowed loops in A stretched to twice as long as the part of the earth's surface shown in Fig. 1, and then compress it until it lies entirely below the level marked C in that diagram.

almost continuously deluges the jungles of the Amazon, Niger and Congo basins, and the islands of Indonesia.

The rising air does not, of course, leave a vacuum behind it, but winds blow in from the north and south to take its place. These are called the *Trade Winds* because in the days of sail merchant ships relied on them for trading in tropical waters. They do not, however, blow directly from the north and south but from the northeast and southeast, having been deflected by the rotation of the earth for reasons given in the next section.

The air rising over the Doldrums cannot go up indefinitely. As it cools and loses its moisture it grows denser, and at a height of about 9 miles it is too heavy to rise any farther. But more is arriving from below, so it moves off to the north and south at this height as the *Anti-trades*. These are subject to the same deflection as the Trade Winds and actually blow over them, but in the opposite directions. As they move they slowly descend, and after traveling about 2,000 miles the northern Anti-trades are sometimes low enough to blow across the Peak of Teneriffe, a volcano in the Canary Islands 12,100 feet high.

This is very nearly the end of their journey, for at about latitude 30°N. they settle down on the surface as a mass of cool heavy air, manifest as a belt of high-pressure calms. These are called the *Calms of Cancer*, and the corresponding belt in the southern hemisphere forms the *Calms of Capricorn*. These are rather loose terms, for the Tropics of Cancer and Capricorn are situated at $23\frac{1}{2}$°N. and S., respectively, but the calms move seasonally with the heat equator and their borders just touch the tropics periodically. In the old days, sailors shipping horses to America dubbed the Calms of Cancer the 'Horse Latitudes,' for if they were becalmed there for many days they were reduced to killing and eating their cargo, a great treat in the circumstances.

The air of these calms is exceedingly thirsty for moisture, for two reasons. First, it dropped most of its rain before it left the Doldrums, and second, as it descended over the Horse Latitudes and fell to warmer and warmer levels, it became able to hold more and more water vapor. This it can take up over the oceans, but over the land it dries up all the water it can find, and this is why there are two belts of deserts round the world at about latitudes 30° N. and S.

On reaching the earth's surface the descending air, being followed by an endless supply from above, again spreads outwards to the north and south. The streams moving towards the equator are the Trade Winds, already described, but those moving towards the poles are deflected by the earth's rotation and become the *Westerly Variables*. In the northern hemisphere they form the prevailing southwest winds that blow across the British Isles. In the southern hemisphere they become strong westerly winds which blow right round the globe at

The Atmosphere and Beyond

about 40°S. Sailors called these the 'Brave West Winds' or, south of the Indian Ocean, the 'Roaring Forties.'

Since the Westerly Variables blow from the tropics their air is warm compared with the masses of cold air over the poles. This cold air, being heavy, spreads slowly outwards and presently meets the tropical air along a boundary called the *Polar Front*. The warm air tends to rise above it but friction along the boundary causes all sorts of disturbing eddies to form. These are the source of the uncertain weather experienced in cool temperate latitudes, and their study belongs to meteorology. The air that succeeds in rising blows round and round the world at a height of about 6 miles as a *jet stream*. See Fig. 19. It keeps over the Polar Front right at the top of the troposphere, and sometimes attains a velocity of 200 miles per hour. It is distinguished as the 'Frontal Jet Stream', for there are other jet streams in subtropical regions although their causes are not yet clear.

Ferrel's Law

That the deflection of the Trade Winds and Westerly Variables is due to the earth's rotation was first suggested by George Hadley in 1735. He gave a simple explanation which is true as far as it goes but is incomplete. It may be illustrated as follows.

Suppose a friend of yours is walking by the roadside and you are overtaking him in a motor car. If you throw a ball *directly* towards him it will land some distance ahead because of the speed of the car. Similarly, if he throws a ball *directly* towards you, as you are passing him, it will land behind the car because all the time it is traveling you are moving away from the point aimed at. Now, if we imagine the ball replaced by a 'piece' or 'mass' of air, so that in traveling it constitutes a wind, we can apply the same principle to explain the deflection of the winds.

All places on the rotating earth are traveling from west

to east, but they do not all move at the same speed. On the equator they travel the full 24,903 miles every 24 hours, which is rather more than 1,000 miles per hour, but at 4 miles from the North Pole, say, they have only about 25 miles to go in the same time, which is very little more than 1 mile per hour. London moves eastwards at about 647 m.p.h., New York at about 755 m.p.h., and places at latitude 30° (N. or S.) at about 866 m.p.h. Winds blowing from this latitude towards the equator are like the ball being thrown from the roadside, for the equator is moving faster. They start off from north to south (in the northern hemisphere) but arrive farther west than they should, so their direction across the earth's surface is from the northeast instead of from the north.

Similarly, winds blowing northward from latitude 30°N. towards New York or London, which are both traveling at slower speeds, are like the ball thrown from the car. They arrive ahead of their expected destinations, and come in from the south-west instead of from the south. Thus, Hadley's explanation may be stated thus: *Winds blowing from places of slow rotation to places of fast rotation lag behind; winds blowing from places of fast rotation to places of slow rotation move ahead.*

But this does not explain why winds blowing exactly due east or west are also deflected, so that other principles must somehow be involved. These were found by William Ferrel, in 1870, to be centrifugal force and the conservation of angular momentum. Since the earth is rotating, any mass of air on its surface is like a stone being swung round on a string attached to the earth's axis. Near the North Pole the string is short and the circles described are small; on the equator it is long and the circles large. Now, the faster a stone is swung round on a string the farther out it flies (by centrifugal force) and the larger the circle it describes. In a similar way, if the air on the surface of the earth starts blowing due east and so increases the speed of its rotation, it will tend

The Atmosphere and Beyond

to fly outward, and this means moving toward the equator where the circles are large. That is to say, it will be deflected toward the southeast. But if the air blows due west, against the rotation of the earth, and thus decreases its speed of rotation, it will seek a smaller circle and move towards the pole, where the 'string' is short. That is to say, it will be deflected toward the northeast.

The conservation of momentum tends to check these effects, causing the fast air moving toward the equator to reduce its speed of rotation as it goes, so that it actually sweeps round in a semi circle, while the slow air moving towards the pole is speeded up and moves in the opposite semi-circle. If we consider the second case first, the air moving towards the poles is having its 'string' continually shortened. Its speed of rotation is then affected exactly like that of an ice skater who, while spinning around on his toes with his arms outstretched, gradually shortens them. He rotates faster and faster until, when his arms are tight against his sides, he is spinning like a top. His total *amount of motion** remains the same, but what he loses in diameter he gains in speed. Similarly, if he spreads his arms again he will spin more slowly, and as the fast-moving air reaches out towards the wider circles near the equator, it loses its speed of rotation for the same reason. These two effects need to be added to Hadley's explanation, but whereas his was a purely kinematic description Ferrel's is a dynamic theory.

Ferrel's complete explanation of the deflection of the winds by the earth's rotation is summed up in 'Ferrel's law,' which is applicable to all objects moving over the earth's surface in any direction, including the ocean currents (as we have seen). It may be stated thus: Objects moving over the earth's surface are continuously deflected to their right in the northern hemisphere, but to their left in the southern hemisphere. Or again, *moving objects tend to*

---

*In Newton's sense, but nowadays called *momentum*.

circle in a clockwise direction in the northern hemisphere, but in a counterclockwise direction in the southern hemisphere.*

Mention of Ferrel's law often raises the question of whether it also determines the direction in which the bath water circles as it runs down the plughole. This is a perfectly good geophysical question, and the answer is that theoretically it should, but practically it does not. One side of the bath – or even one end of the bath – does not move significantly faster on the rotating earth than the other side or end, and the almost infinitesimal difference is quite negligible in the face of such other influences as an accidental swirl of the water, a slight off-center position of the plug hole, the exact tilt of the bath, the flick of the plug as it is withdrawn, or even the lay of the drainpipe in the trap underneath. Any particular bath, or perhaps any particular make of bath, is liable to be biased in favor of one direction of rotation, but it would be a mistake to attribute this constancy to the rotation of the earth. In the author's bathroom, undisturbed water always empties in a clockwise direction from the bath, but in a counterclockwise direction from the sink.

Further Modifications

We have outlined the ideal *planetary wind system* (first mapped by Edmund Halley in 1688), and have shown how the directions of the permanent winds are modified by the earth's rotation. There are, however, other modifications of a major character about which something should be said. These are caused by the great land masses of the continents, which are separated by even larger oceans.

---

*This is often confused with Buys Ballot's 'law' of 1857, which merely gave ships' captains a working rule (based on the deflection of the winds) for dodging low-pressure weather systems. It stated that if you stand with your back to the wind, the area of low pressure lies to your left in the northern hemisphere, but to your right in the southern.

The Atmosphere and Beyond

The principle involved is that, under the sun, the land both warms up and cools down much more quickly than the sea.

In the summer the air over a large continent becomes hotter, and in winter colder, than the air over the surrounding oceans. In both cases, the air over the warmer region is at lower pressure than the air over the colder region, and winds tend to blow from the colder areas into the warmer. Normally, this effect is strong enough only to modify the permanent winds. In summer the Trade Winds, for example, are pulled more sharply into the continents along their east coasts, so that they come in from almost due east, while on the west coast they do not leave the continent at the expected sharp angle because they are partly held back. In winter they are checked or accelerated in the reverse manner.

These effects are greatly exaggerated over Asia, which is so vast that it becomes hot enough in the summer to draw winds into it from all sides, willy-nilly. It even draws the southeast Trade Winds right across the equator, where they change their direction according to Ferrel's law and cross India as the southwest *monsoon*. This is India's wet monsoon, and it drops heavy rain when it is forced upwards by the mountains of the Western Ghats. On approaching the snow-capped Himalayas it rises 5 miles and gives the heaviest rainfall in the world to Nepal and Assam. This is about 40 times more rain than London receives in a year, and 25 times more than that of New York.

In winter, central Asia becomes a vast pool of cold dense air which spreads outwards in all directions, pouring over India as the dry north-east monsoon. Similar monsoons are drawn in over China during the summer and blown out again in the winter, but this air comes all the way from Australia. The details may be studied in the maps of the permanent winds for January and July given in almost every atlas of the world.

Exploration Upward

The most important general characteristics of the atmosphere are its pressure, temperature, composition, and electrical state at different levels. The study of these began with the measurement of pressure when the Italian physicist Evangelista Torricelli invented the barometer in 1643, and the French mathematician and philosopher Blaise Pascal showed, in 1646, that the pressure decreases with altitude. It has been found that the pressure near sea level drops by about half an ounce per square inch for every 60 feet of ascent, but since air is compressible it rapidly becomes thinner (*i.e.* less dense) as the pressure is reduced. At 18,000 feet above sea level the drop is only a quarter of an ounce per square foot for every 60 feet, but the rate is also greatly affected by the temperature. At the top of the troposphere the pressure is only about 4 pounds per square inch and the air is much too thin to breathe.

Exploration of the atmosphere upwards was begun by climbing mountains, but this is arduous and in 1749 Alexander Wilson began sending thermometers aloft on kites. In 1893, F. Bezançon and C. Hermite used small balloons carrying self-recording instruments, which were later combined in a unit called a *meteorograph*. These are known as *ballons-sondes* or 'sounding balloons'. As they ascend into the rarefied air the gas within them expands until, at a roughly predetermined height, the balloon bursts and the meteorograph returns to earth by parachute. Sounding balloons have been sent to very great heights, the practicable limit of about 25 miles being attained just before World War II.

Sounding balloons have now been largely replaced by *radio-sondes*, invented by P. Idrac and R. Bureau in 1927. These register the pressure, temperature, humidity and other characteristics of the atmosphere as they ascend, and broadcast the data continuously. The record is thus provided without any delay, and it is of no consequence

The Atmosphere and Beyond 95

Fig. 16. A radio-sonde rising steadily shortly after its release. (*French Embassy*)

if the instruments are never recovered. See Fig. 16. In 1966, the first of a series of 'GHOST' (Global Horizontal Sounding Technique) balloons was released, and these remain permanently floating at great heights in the atmosphere, where their motions can be followed by radar and their reflection of radio waves studied. Figs. 19 and 20 show the main phenomena of the atmosphere that have been discovered up to a height of 350 miles.

The Stratosphere

The early sounding balloons yielded the expected information that, even away from snow-capped mountains, the temperature of the air steadily drops with increasing height. But in 1899, Teisserenc de Bort and R. Assmann made the astonishing discovery that above a height of 7 or 8 miles the temperature ceases to fall and winds cease to blow. The air here contains no appreciable water vapor and no normal clouds, but remains static up to a height of about 30 miles. This layer of the atmosphere is called the *stratosphere* (meaning 'layer sphere'), and its ill-defined junction with the troposphere, where the jet-streams blow, is called the *tropopause*. See Fig. 19.

The temperature of the stratosphere, which is about −80°F in the bottom layers, begins to increase again above 15 miles, reaching the freezing-point of water at about 25 miles. At a height of about 20 miles faint luminous clouds probably consisting of supercooled water droplets are occasionally seen at dusk and far into the night. Known as *iridescent* or 'mother-of-pearl' clouds, they show an ever-changing pattern of rainbow colors and provide one of the most beautiful displays in the sky.

As information about the stratosphere began to accumulate, scientists thought a personal visit desirable, and in 1931 Auguste Piccard and Paul Kipfer rose to a height of 10 miles in a sealed aluminum ball slung beneath a hydrogen balloon. Other ascents followed, helium being generally preferred to hydrogen because it is non-inflammable. The record altitude (in 1968) stood at $23\frac{1}{2}$ miles and was attained by N. Pianstanida over South Dakota in 1965. The pressure of the atmosphere at that height is about $2\frac{1}{2}$ ounces per square inch, or only one-hundredth of the pressure on the earth's surface. Fig. 17 shows the start of an earlier record ascent, in which 17 miles was reached.

Since World War II much greater heights have been

The Atmosphere and Beyond

Fig. 17. The ascent of the U.S. third *Manhigh* stratosphere balloon in New Mexico, 1958, in which Clifton McClure remained aloft at 99,600 feet (17 miles) for 24 hours. In the near-vacuum of high altitudes the gas expands and fills the envelope, which is here seen collapsed under normal atmospheric pressure. (*U.S.I.S.*)

Fig. 18. Launching a *Skylark* upper-atmosphere sounding rocket. The *Skylark*, which is 25 feet long and burns solid fuel, carries about 125 pounds of geophysical instruments to a height of 100 miles. (*British Aircraft Corp.*)

The Atmosphere and Beyond

reached by rockets. The liquid-fuel 'V-2,' used to bombard London, rose 60 or 70 miles and, after the war, was used for research in the upper atmosphere to heights approaching 120 miles. More than sixty of them carried a total of 20 tons of scientific instruments well beyond the stratosphere before cheaper rockets were developed. Most important of these was the *Aerobee*, which played a big part in the International Geophysical Year program. The *Skylark* (Fig. 18) was a pioneer in solid-fuel rockets, but it was still very expensive for research purposes and much smaller rockets, such as *Deacon*, *Terrapin*, *Cajun* and *Kappa*, were devised. These rise only about 10 miles if fired from the ground, but if they are first carried 15 miles up by a balloon and then fired, they will take instruments up to 60 miles or more. This combination is known as a '*rockoon*.'

Instruments sent to such heights have shown that, above about 30 miles, the temperature begins to fall again, dropping fairly quickly to $-100°F$. at a height of about 50 miles. This zone of the atmosphere is called the *mesosphere* (meaning 'middle sphere'), and it is succeeded by the *thermosphere* ('heat sphere') in which the temperature once more rises – and continues to rise as far as the atmosphere can be detected. It reaches freezing point at about 65 miles and boiling point ($212°F$.) at 75 miles, continuing to rise until it attains more than 4,000 degrees. See Fig. 20.

If you visited these regions you would not feel warm, however, for although the temperature is high the *amount* of heat is almost negligible. You can pass your finger quickly through a candle flame without feeling it for the same reason – the temperature is high but you do not receive sufficient heat for it to be effective. Temperature is a measure of the speed at which the molecules are rushing about and colliding, and at a height of 100 miles, say, there is only one molecule or atom for every 250 million in the air you breathe on the ground. You would not even

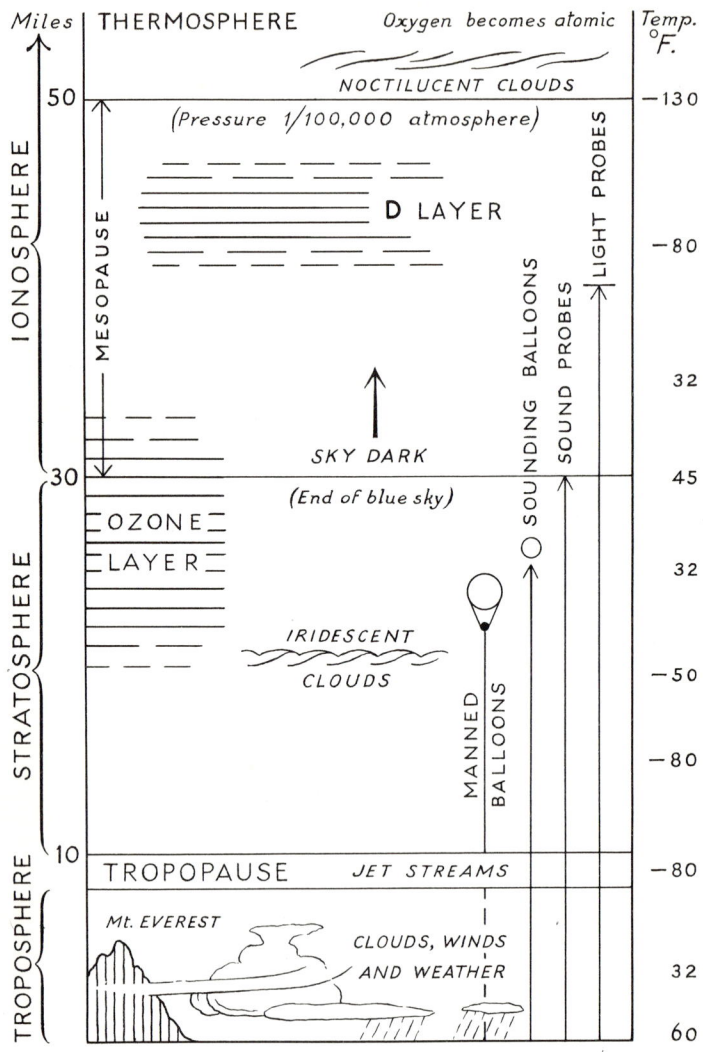

Fig. 19. Height profile of the atmosphere up to 50 miles. The sky becomes dark as the density of the air falls below 1/2,500 of the density at the surface, though it still contains some 100,000 million million molecules per cubic centimetre.

The Atmosphere and Beyond

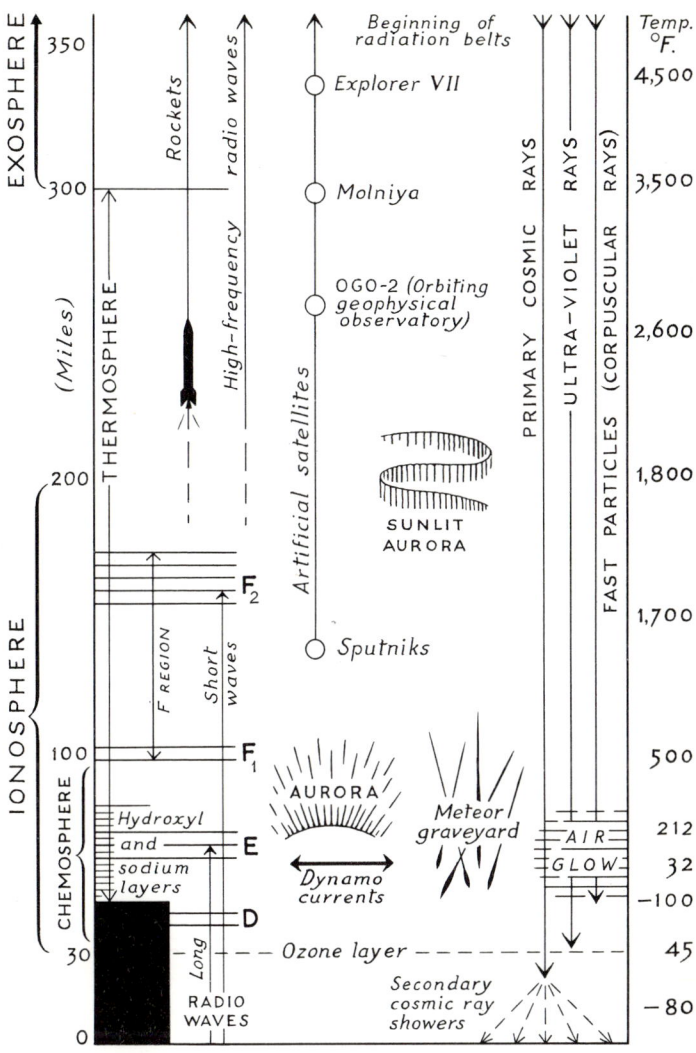

Fig. 20. Height profile of the atmosphere up to 350 miles. The details covered by the black rectangle are shown on a larger scale in Fig. 19. Note the large parade of phenomena in the chemosphere, about 70 miles above the ground.

know there were any there at all.

At the base of the thermosphere (50 miles up) another mysterious kind of cloud sometimes appears. Called *noctilucent* clouds because they are seen shining as night approaches, they form silvery blue wisps and tissues, often billowing like ghostly waves. They are seen only at high latitudes and in the summer, and evidently reflect light from the sun below the horizon. Their nature remained a mystery until 1960, when a *Nike-Cajun* rocket carrying particle collectors was sent up in Sweden to bring back specimens of the clouds' material. This turns out to consist of microscopic solid particles, seen to be perfectly spherical when magnified 30,000 diameters in an electron microscope. They appear to contain nickel and are probably derived from burned-out meteors, but when in the cloud they are possibly coated with ice.

Probing with Waves

There are other ways of exploring the upper atmosphere besides sending up instruments. During the Dutch War of 1666, Samuel Pepys, the diarist, noticed a 'miraculous thing.' He observed that the gunfire in the English Channel could be heard in St. James's Park, London, and yet was inaudible on the coast at Deal and Dover. The same phenomenon was observed during World War I, but long before that the correct explanation had been found. It was that the sound waves traveling over the surface of the earth are rapidly absorbed by hills, houses and trees, and soon die away, but those that travel upwards at an angle continue until they reach a layer of warm air which reflects them down again to places very much more distant.

To ascertain the exact height of this warm layer, arrangements were made in 1927 to record the arrival of sound waves from gunfire in the mouth of the Thames at Nottingham, Sheffield, Birmingham, Bristol and Cardiff,

The Atmosphere and Beyond

by means of sensitive microphones. Radio signals gave the times of firing, and from the results it was calculated that the sound waves left the muzzles of the guns in a gigantic upward curve until they reached a height of about 9 miles. This took about a minute, and the curve reflected the drop of temperature through the troposphere. For the next 15 miles the waves traveled in a straight line through the nearly even temperature of the stratosphere, but then they began to curve over and at about 30 miles were reflected downwards again, confirming the rise of temperature in the upper stratosphere.

The cause was deduced from spectroscopic observations made by G. M. Dobson, who had noted the presence of a belt of ozone about 25 miles above the earth. The ozone is produced by the action of ultra-violet rays from the sun on the oxygen in the atmosphere, and this involves the trapping of sufficient heat to account for the rise in temperature. The quantity of ozone in the belt, which is about 10 miles thick, is very small because of the rarity of the atmospheric gases at that height. If it could be brought down to the surface of the earth, it would make a layer only about 3 millimeters thick. Nevertheless, life on the earth depends upon it to screen off all but a tolerable sprinkling of the dangerous ultra-violet rays. Local concentrations of ozone also seem to have some influence on the weather far below in the troposphere, but this is not yet fully understood.

Light waves have also been used to explore the atmosphere beyond the range of balloons. A powerful searchlight will project a beam more than 30 miles, but its light is to some extent scattered by the molecules or atoms of the air. The scattered light from a small selected area at this height is much too faint to be seen, but it can be collected by another searchlight mirror and focused onto a photomultiplier, which enables its intensity to be calculated. This, in turn, reveals the density of the scattering particles, which is the density of the air at that height.

The Ionosphere

The most useful waves for probing the upper atmosphere are radio waves. Radio waves, like light waves, travel in straight lines. When G. Marconi sent a wireless message across the Atlantic Ocean for the first time in 1901, it was not known how the radio waves had managed to bend round the curvature of the earth. Oliver Heaviside had already suggested, in 1892, that the short-wave radiation from the sun probably ionizes the atoms in the upper atmosphere. He now proposed that the radio waves crossed the Atlantic by being reflected down from a layer of ionized gas at a height of about 65 miles. The same idea occurred independently to A. E. Kennelly, and this layer, whose existence was proved by E. Appleton in 1924, is now called the Kennelly-Heaviside layer. The solar radiation causing it is composed mainly of X rays, which are energetic enough not only to break up the molecules of oxygen into seperate atoms, but also to strip the outer electrons off the atoms or *ionize* them. The free electrons render the atmosphere conducting, and it can be shown that a conducting layer will reflect radio waves.

The Kennelly-Heaviside layer reflects only long and medium radio waves. Short radio waves pass right through it, though they also may be received round the curvature of the earth. This problem was cleared up in 1926 by E. Appleton, who discovered another reflecting layer of ionized gas at a height of about 140 miles. Since then other layers of high electron concentration have been found, and they are usually referred to by letters. See Fig. 20.

The lowest of these layers occurs at from 40 to 50 miles, but only at times of exceptional outbursts of radiation from the sun. Known as the D layer, it is responsible for radio fade-outs and other interference with broadcasting. Next comes the Kennelly-Heaviside layer, which is

The Atmosphere and Beyond

denoted by E and is about 6 miles thick. Above this comes the Appleton layer, now known as the F 'region' because it divides into two distinct layers during the daytime; these are the F_1 layer at a height of about 100 miles, and the F_2 layer at about 160 miles. The F_2 layer is approximately 30 miles thick and its temperature may be equivalent to about 1,200°F. The whole zone of the atmosphere containing these layers is referred to as the *ionosphere*, and it lies wholly within the thermosphere. Only very high frequency radio waves can pass right through it, so that these have to be used for signaling to artificial satellites.

The Chemosphere

Another zone, depending upon different characteristics, is known as the *sodium layer*. This probably begins at a height of about 45 miles and extends upwards to 60 miles or more (Fig. 20). It is responsible for a very faint luminescence visible on clear dark nights as the 'night air glow', which appears to be due chiefly to the recombination into molecules of atoms separated and ionized by the sun's radiation during the day. These molecules are evidently fluorescent, radiating visible light as they settle down from an excited state. The spectroscope shows that the principal atoms responsible are those of sodium, though it is difficult to account for sodium at such a height unless it is supplied from burned-out meteors. However, oxygen also plays a part. The strong yellow sodium line, a fainter red oxygen line, and a blue-violet line caused by molecular nitrogen are responsible for the momentary enhanced glow known as the 'twilight flash,' which is sometimes visible soon after the sun has set.

In the same zone lines in the infra-red part of the spectrum, formerly attributed to nitrogen, are now known to be emitted by hydroxyl ions (OH), formed by the separation of some of the hydrogen from water molecules, so that a *hydroxyl layer* is also sometimes referred to. But the whole region from the ozone layer to a height of about 100 miles contains several unusual chemical groups and has been called the *chemosphere*.

By measuring the width of the yellow sodium line, it is possible to estimate the temperature of the sodium layer. This has led to the artificial dissemination of sodium vapor at other levels in the upper atmosphere for temperature determinations. The first successful experiment was made in 1955, when two containers, each holding 2 pounds of metallic sodium with a 'thermite' mixture to vaporize it, were carried aloft on a rocket shortly after sunset. The containers were fired at heights of 42 and 47 miles, respectively, and the discharged sodium vapor formed a luminuous cloud extending from 53 to 70 miles above the earth. It was visible to the naked eye from as far away as 300 miles, and its rapid distortion showed that winds of some kind were blowing at those levels.

In similar experiments performed in 1956, large volumes of nitric oxide (the gas NO) were discharged from a rocket at heights of 58 and 65 miles. The result was a strongly luminous pale green glow about 16 times the size of the moon, with a measured total brightness of about a million candle power. The phenomenon indicated that the atmospheric oxygen in that zone is mainly in the form of atoms instead of molecules. The motions of the air were also found by tracking the cloud with radar. Numerous other experiments of similar kinds have since been performed, and it appears that, unlike the oxygen, the atmospheric nitrogen remains in the form of molecules to very great heights indeed.

The Exosphere

The *exosphere* is the outermost layer of the atmosphere, but the gas composing it is so thin that it would be

The Atmosphere and Beyond

regarded as a high vacuum in the laboratory. The molecules of any gas are always in rapid motion, colliding with each other and going off in new directions in a completely random manner. In the air at the earth's surface they travel at an average speed of about 1,500 m.p.h., which is about the speed of a rifle bullet, but they are so small and so close together that each suffers some 3,000 million collisions per second. The exosphere is defined as that region in which the speed of at least some of the particles exceeds the earth's escape velocity of 25,000 m.p.h., and where they are so rare that many are able to travel a distance equal to the earth's radius (3,957 miles) without a single collision.

The exosphere (Fig. 20) is assumed to extend indefinitely upwards from about 300 or 400 miles above the earth's surface, but its temperature, and therefore the speed of its particles, is not definitely known. A conservative estimate of 3,600°F. would give a mean speed of 16,700 m.p.h. for the fastest kind of particles (hydrogen atoms), but this means that *some* will undoubtedly exceed to escape velocity and fly off into space. The exosphere is thus the region from which the earth's upper atmosphere is slowly leaking away, although only the two lightest gases (hydrogen and helium) have diminished appreciably in this way since the earth was formed.

The composition of the atmosphere in the exosphere, and its density, are no more exactly known than its temperature, though a rough idea has been gleaned from the movements of artificial satellites through it. Though the air is extremely thin it provides sufficient friction to slow down a satellite to a very minute degree, and this causes its orbit to become slightly more circular. The differences are so small that they can be detected only after several hundred orbits have been made, but the results give an approximate measure of the density. From this, and the probable temperature, the mean molecular or particle weight of the air can be roughly estimated, and its probable composition can be deduced from this. It appears that the exosphere consists largely of hydrogen, with some helium and very little oxygen and nitrogen.

Artificial satellites perform a great variety of other duties. These include the estimation of meteoric dust in and above the exosphere, the measurement of cosmic rays and solar radiation, the evaluation of electron densities, and the study of electric and magnetic fields. The meteors and cosmic rays are visitors to the earth from space and are dealt with in astronomy books, but their influence on the earth's atmosphere is a geophysical question and they are included in Fig. 20.

Meteors and Cosmic Rays

The majority of meteors are of microscopic size, but those as large as grains of sand become visibly incandescent by friction with the air and are seen as 'shooting stars'. They fall at about 80,000 m.p.h., which is fast enough to melt and vaporize them even in the thin air above the stratosphere. Most meet their fate between 100 miles and 70 miles above the ground, this zone being known as the 'meteor graveyard'. Since they leave long trails of ionized air behind them they can be detected by radar, and the reflected radio waves provide information about the temperature and pressure at that height.

The cosmic rays are of two main kinds, the original or *primary* rays consisting of highly energetic particles from an unknown source in space. These are usually protons or hydrogen nuclei, but the nuclei of much heavier atoms are also present. When they collide with the atoms of the atmosphere they produce showers of *secondary* rays, which include mesons, electrons and gamma-rays. Many of these not only reach the surface of the earth, but penetrate deeply into the rocks and the sea. On the average, the cosmic rays disintegrate about 20 atoms in every cubic inch of the atmosphere, and many millions

The Atmosphere and Beyond

of atoms in our bodies, every second. Their main effect in the upper atmosphere is to maintain the continual release of free electrons by ionization.

Dynamo Currents

The ionosphere is characterized by the relative abundance of free electrons, especially in the highly ionized layers already described. Movements of these electrons constitute electric currents and hence generate magnetic fields. These were first observed as unexplained fluctuations in the earth's magnetic field at ground level, and when they were strong enough to disturb ships' compass needles they were called 'magnetic storms.' Close examination showed that they exhibit periods corresponding with those of the tides, and they were attributed to currents of electrons in the ionosphere, caused by tides in the atmosphere similar to those in the ocean. Unlike ocean tides, however, the effects of the sun are greater than those of the moon, for the tidal bulge caused by the sun's gravitational pull is augmented by thermal expansion due to the sun's heat.

This theory was originally suggested by Balfour Stewart in 1882, and was worked out in detail by A. Schuster in 1908. In effect, the atmospheric tides cause streams of ionized air to move across the lines of force of the earth's magnetic field. In doing so, they behave like the moving coils in a dynamo and generate the strong currents that cause the observed disturbances on the ground. A particularly intense dynamo current flowing at from 50 to 70 miles above the equator is known as the *electrojet*.

The Magnetosphere

The earth's magnetic field has been detected far out in space, and the directions of its lines of force have been conjectured. Though they are by no means symmetrically

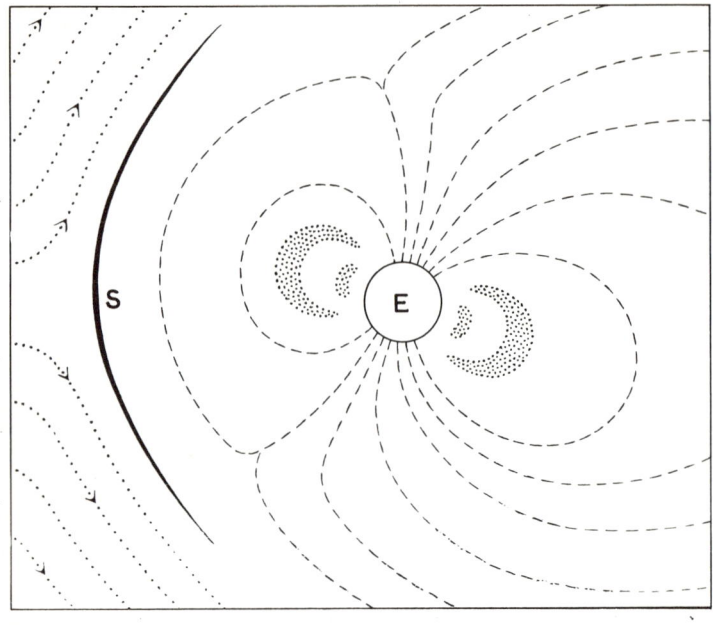

Fig. 21. The magnetosphere. E, Earth. S, Shock front. Broken lines: lines of force of the earth's magnetic field. Dotted lines: Solar wind. Dotted areas: Van Allen radiation belts.

disposed around the earth, the volume which they occupy in appreciable strength is called the *magnetosphere*. See Fig. 21. This is the field generated by the currents in the liquid outer core of the earth, as described in Chapter 3, and its first effect is to provide the earth with two magnetic poles, like those of a bar magnet. They do not coincide with the geographical poles, and neither are they exactly opposite each other. Further, they drift about so that compass readings have to be corrected according to the 'magnetic variation' given in almanacs. More astonishing is the periodic reversal of the poles, north becoming south and south north, every 450,000 years or so. The evidence for this is given on page 46, but the reasons for it are quite unknown.

The Atmosphere and Beyond

The distribution of the earth's magnetic field in space has been found chiefly by artificial satellites, of which an example is shown in Fig. 22. Theoretically, it should form a kind of huge doughnut surrounding the earth, with the poles running through the doughnut's 'hole', but it turns out to be enormously distorted by a continuous 'wind' of protons and electrons, which are discharged from the sun and travel out far beyond the earth's orbit. This *solar wind* constitutes an electric current generating its own magnetic field, and this compresses the magnetosphere on the sunward side of the earth, where it extends to about 40,000 miles, and draws it out in a tail perhaps 3 million miles long on the dark side.

Normally, the solar wind travels at about 670,000 m.p.h., and contains about 170 particles per cubic inch. As it approaches the earth's magnetosphere it is abruptly checked and forms a sort of 'shock wave' at a distance of about 60,000 miles, much like the shock wave produced in the air at the front of a supersonic aircraft. Large quantities of particles are caught in the earth's magnetic

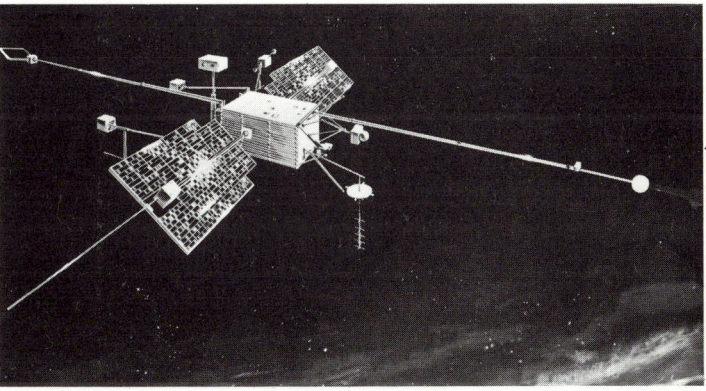

Fig. 22. One of the *OGO* (*O*rbiting *G*eophysical *O*bservatory) satellites, which weigh about half a ton and carry two hundredweight of scientific instruments. Note the large rectangular arrays of solar cells for converting the sun's radiant energy into electricity. (*U.S.I.S.*)

field, but the main stream divides and flows round the earth like a river round a boulder, producing a turbulent wake or slip-stream. It is possible that the *zodiacal light*, a faintly luminous tongue-shaped area of sky sometimes seen after sunset or before dawn, is the reflection of sunlight from particles in the slipstream.

The first satellites designed to explore the magnetosphere were *Pioneer I* (1958), *Vanguard III* and *Explorer VI* (1959), *Pioneer V* (1960), and *Explorer X* (1961). There have been several others since, but the first important discovery was not made by any of these, but by *Explorer I* (1958)—through the failure of its instruments! It carried particle detectors which broke down at a height of about 500 miles, and that might have been the end of its story had not James Van Allen realized just *why* they had broken down. He saw that they had merely been swamped with more electrons and protons than they could deal with. The satellite had, in fact, run into a belt of captured particles of high density, which was later found to be about 2,000 miles thick. Another belt of exceptionally high density, at least 4,000 miles thick and consisting chiefly of electrons, was then discovered at a height of 15,000 miles, and they are now called the *Van Allen radiation belts*. See Fig. 21.

The inner belt forms a fat band round the earth's intertropical regions, with its base about 400 miles up in the exosphere. The outer belt is horseshoe-shaped in section and encloses the inner belt, its northern and southern edges reaching to latitudes 60°N. and S., over which they stand at a height of about 200 miles. These shapes are determined by the lines of force of the earth's magnetic field, for the particles concentrated in the belts are held there in a magnetic trap.

On arriving in the solar wind, the particles are attracted to the magnetic poles and travel along the curves of the horseshoe. See Fig. 23. They move in corkscrew-like paths until they converge near the poles, when they are

The Atmosphere and Beyond

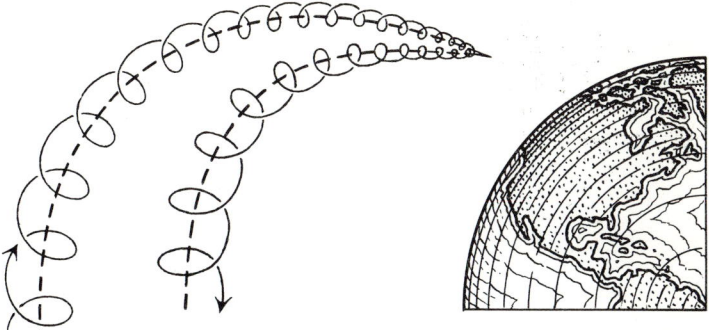

Fig. 23. How the particles from the solar wind are trapped in the earth's magnetic field to form the Van Allen radiation belts. The broken lines represent the horseshoe-shaped magnetic lines of force, and the electrons follow the spiral paths back and forth between the north end (shown here) and the south end. The protons follow similar paths but spiral in the opposite directions.

reflected back again and travel round to the opposite poles. They thus go back and forth indefinitely, only a few colliding with molecules in the atmosphere at the lower levels. Those that do this produce X rays which have been detected with instruments carried by balloons.

The Aurora

Normally, the solar wind takes about 6 days to reach us from the sun. However, at times of great solar activity a blast from a solar flare may arrive in a little less than a day, the particles being finally accelerated to enormous speeds in the neighborhood of the earth. They penetrate the atmosphere to a height of only 70 miles, which means that the protons must have been traveling at about 20 million m.p.h., and the electrons at 200 million m.p.h! They arrive in a deluge which circles the polar regions in *ring currents* of about 300,000 amperes at a million volts or more, and cause severe magnetic storms. Known as the *polar electrojets*, they spray the entire ionosphere with electrons.

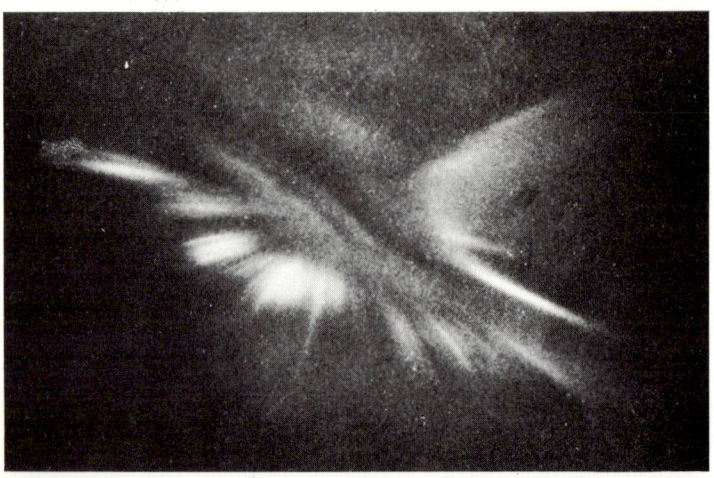

Fig. 24. *Top:* Auroral crown (corona), Norway, 1920. *Below:* Curtain aurora, Norway, 1926.

The Atmosphere and Beyond

The collisions that occur with the atmospheric gases produce spectacular effects. The protons collide with oxygen atoms and nitrogen molecules, raising them to excited states from which they collapse with the emission of light rays. At the same time the electrons are captured by some of the ionized particles and more light is emitted, the total effect making a magnificent display in the night sky. This is known as the 'aurora borealis' in the northern hemisphere, and the 'aurora australis' in the southern, or more simply as the 'polar lights'. See Fig. 24. Such displays most often occur in fairly narrow belts round the magnetic poles centered over latitudes 66°N. and S., but are so high – up to 100 miles – that they can be seen from distances of several hundred miles.

The auroras themselves exhibit two basic forms. They appear either as diffuse cloudlike patches, sometimes with radiating streamers which may form an 'auroral crown,' or as some form of ribbon or curtain of light, which curves and waves in constant motion. Such a curtain may be 200 miles wide and 3,000 miles long, and hang right across the sky, or it may be a looped and folded ribbon with ripples in the folds, when it is called a 'rayed band'. The light may be of several colors, the lower edge usually being red or pink, the middle pearly white, and the upper part blue, green or violet. There are constant flickerings and fluctuations in intensity. A strong aurora may be as bright as the full moon, and displays have occasionally been seen at dawn or even in late afternoon daylight.

Another luminous effect, called the *sunlit aurora*, occurs up to a height of about 200 miles. These lights resemble the others but are blue-green and somewhat fainter. They are caused mainly by collisions that ionize nitrogen molecules, the nitrogen ions being then rendered fluorescent by sunlight falling on them. The heights of the auroras are shown in Fig. 20.

The various phenomena of the upper atmosphere are

still understood only in outline, and are explained by theories which are constantly being tested by experiments with rockets, radio waves and artificial satellites. More remains to be discovered than has yet been found, but we may well marvel at the skill with which the instruments of detection have been devised and the subtle clues they provide interpreted.

APPENDIX

The Biosphere

The parts played by living organisms in determining the geophysics of the earth as we know it today have not been detailed in this book, since they belong mainly to the remote past. They form chapters in the stories of historical geology and climatology, and run right back to the beginnings of life on the planet. Nevertheless, they are of prime importance as causes, and a balanced picture of the world we live in demands their recognition in a few brief paragraphs.

The earth's present atmosphere, for example, owes its composition chiefly to the emergence of plant life from the sea in Silurian times, perhaps 400 million years ago. At that time, and throughout the previous 4,000 million years, there was little or no oxygen in the air. Thus, for some 90 per cent of the earth's long story the atmosphere has been unbreathable and quite possibly poisonous.

When the crust of the earth first solidified the atmosphere probably consisted chiefly of nitrogen, hydrogen, argon, carbon dioxide and water vapor, with a fair amount of ozone, ammonia, methane, and perhaps cyanogen manufactured by X-rays and ultra-violet radiation from the sun, and by gamma-rays from radioactive elements. When the surface was cool enough to support water, the ammonia and other soluble gases dissolved, and by about 1,000 million years ago it is thought that the atmosphere had much the same composition as today, except for the virtual absence of oxygen and an excess of carbon dioxide. If there was any oxygen, it was probably derived from molecules of water vapor or hydroxyl ions (OH), broken up by the action of the sun's radiation high in the atmosphere, where the hydrogen could escape before being recaptured.

The advent of living matter in the seas eventually led to green plants, which are able to manufacture sugars from carbon dioxide and water with the help of chlorophyll, a complex green compound containing magnesium. The energy required to build up the sugars is obtained from sunlight, and oxygen is evolved as a waste product. This would have dissolved in the water until the plants began to invade the land, but when this happened it would be released as a gas in exchange for the carbon dioxide consumed. After several hundred million years the atmosphere was completely transformed, and today there is abundant oxygen for the support of air-breathing animals while the carbon dioxide has been reduced to little more than a trace. (See Table V, page 85.)

Before this great change took place the rocks were probably mainly gray or black, like those of the moon, but the presence of free oxygen led to their oxidation and the production of the reds and yellows characteristic of the oxides of iron associated with rust. Attempts have been made to fix the period of the change by dating the oldest known reddish sandstones, but since the process must have been gradual, and changes may have taken place in some of the sandstones since they were formed, there have been no very informative results. Very old river gravels have, however, been found containing crystals of iron pyrites (iron sulphide) which would certainly have rusted away beneath an atmosphere containing much less oxygen than the one we breathe.

The advent of land vegetation must also have modified the climate over vast areas, lowering the temperature range, increasing the humidity, and sheltering the surface while at the same time breaking it down into soil. In some

Appendix 119

regions the accumulated dead vegetation eventually became coal, which occurs in sufficiently massive deposits to be called a rock and to be considered as a structural element in the lithosphere.

Coal is only one of several organic rocks, the chief of which is limestone. Enormous areas of the earth's surface consist of beds of limestone, sometimes measuring several thousand feet in thickness. Limestones are sometimes deposited by chemical precipitation but the majority are accumulations of organic remains, consolidated by pressure and partial solution. More than 1,000 million years ago lime-secreting seaweeds built up enormous reefs which later formed limestones. Coral reefs make similar limestone formations, and so do the remains of submarine forests of sea lilies, and even the shell-banks where dead seashells have accumulated under the action of currents. Marine bacteria are known to be precipitating a limy ooze off the Bahamas, while other limestones, such as chalk, are formed from the microscopic shells of coccoliths and Foraminifera.

Of all these things you may read at length in books on geology and physical geography. They come into geophysics chiefly as results—as *faits accomplis*. They modify the density of the outer crust of the earth, they act as weights in the scales balanced by isostasy (page 37), and they are often responsible for the appearance of dry land where there would otherwise have been sea.

Man himself may also be considered as a geophysical agent, for he has already altered climates by land reclamation, re-afforestation, drainage, irrigation and dam building, and he has learned how to modify the weather in some circumstances, producing artificial rain and snow storms. His invasion of space has already produced local electrical disturbances in the upper atmosphere, and his experiments have provided the earth with a new family of satellites. He may one day change the entire face of the planet by such projects as melting all the polar ice by nuclear heat, flooding and irrigating the Sahara, deflecting the Gulf Stream to flow up the east coast of America, or by providing continuous night illumination by means of a brilliant artificial aurora of continental dimensions.

Index

Index

Abyssal plain, 58, 59, 61, 63
Aerobee, 99
Airy, George, 36, 37
Alexander the Great, 76
Allen, James Van, 112
Aluminaut, 80, 81
Aluminium, 31, 36–37
Alvin, 80
Amphidromic point, 72
Anaxagoras, 16
Angara Land, 38
Antarctic, 9 ff., 40, 66
  Circle, 87
Anti-trades, 87, 88
Appleton, E., 104
Appleton layer, 105
Archimède, 83
Arctic, 9 ff., 66
  Circle, 87
'Arctis, 38
Aristarchus, 16
Aristotle, 16, 75
Assmann, R., 96
Atlantic, 28, 38, 39, 40, 59, 61, 64–65, 66, 70, 72, 104
Atmosphere, 84 ff.
  circulation of, 86 ff., 108
  composition of, 85. 107
  height of, 18, 19, 107
  primitive, 117–118
  profiles of, 100–101
Aurora, 11, 101, 113 ff., 120
  sunlit, 101, 115

Bacchus, 79
Bacon, Francis, 8, 39
Ballons-sondes, 94
Balloon, sounding, 11, 100, 113
  stratosphere, 96, 97, 100
Barton, O., 81
Basalt, 37, 54, 55, 58
Bathyscaphe, 82–83
Bathysphere, 81, 82
Beche, Henry De la, 33
Beebe, C. W., 81
'Bends', the, 76
Bezançon, F., 94
Biosphere, 20–21, 117 ff.
Bismarck, Karl Otto von, 9
Blackett, P. M. S., 45
Bond, G., 78
Bore, 73
Bore-hole (boring), 26, 53–54
Borman, Frank, 20
Bort, Teisserenc de, 96
Brahe, Tycho, 18
Brave West Winds, 87, 89
Bullard, E., 74
Bureau, R., 94
Buys Ballot's law, 92 *n.*

Calcium, 31, 37, 63
Cancer, Calms of, 88
  Tropic of, 86
Capricorn, Calms of, 88
  Tropic of, 86
Carbon dioxide, 85, 117, 118

Geophysics

Cavendish, Henry, 18, 38
Challenger, H.M.S., 62, 74
Chandler wobble, 42 *n*.
Chemosphere, 101, 105–106
Clairaut, A. C., 22
Clouds, 96, 100
    iridescent, 96, 100
    noctilucent, 100, 102
    sodium, 106
Coal, 38–39, 119
Cobalt, 31, 61
Conductivity, electrical, 14, 104, 109, 113
    thermal, 27
Conshelf, 78, 79
Continental block, 37, 39, 40, 45, 47, 54, 55
    drift, 39 ff., 52
    edge, 58
    platform, 58
    rise, 58
    shelf, 37, 58, 59, 61, 62, 63, 79, 80, 81
    slope, 58, 59
Continents, 36 ff., 44 ff., 47, 52, 92 ff. *See also under* Continental block
Convection, 42
    in atmosphere, 86 ff.
    in oceans, 64 ff.
    in rocks, 42 ff., 52, 54–55, 59
Copernicus, Nicolaus, 17
Corona, auroral, 114
Corpuscular rays, 101. *See* Electrons, Protons
Cosmic rays, 101, 108–109
Cousteau, J. Y., 77, 78
Creer, K. M., 47
Cromwell, Townsend, 66
Current, convection, *see* Convection
    electric, 52, 109 ff.
    ocean, 64 ff.
    ring, 113
    submarine, 66, 74–75
    tidal, 72–73

D layer, 100, 101, 104
Dana, James Dwight, 34
Darwin, Charles, 28
Dating, radioactive, 28–29, 47
Daubrée, Auguste, 39, 43
Davis, Robert, 76
Decompression, 76 ff.
Deeps, 57, 58, 81–83
Density of air, 84, 103, 107
    of earth, 38, 119
    of sea-water, 84
Descartes, René, 22, 26
Discontinuity, 50
    Gutenberg (Oldham) 52
    Mohorovičić, 37, 52, 53
Discovery II, 28
Diving, 75 ff.
Dobson, G. M., 103
Doldrums, 86, 87, 88
Dutton, C. E., 37
Dynamo currents, 101, 109

E layer, 101, 105
Earth, age of, 28–29
    appearance of, 20, 21
    axis of, 16, 17, 42
    composition of, 29 ff.
    core of, 38, 50–52, 110
    crust of, 26, 28, 30–32, 34, 37, 39 ff., 51, 52, 57, 59, 61, 119
    curvature of, 19, 104
    density of, 37–38, 48
    expansion of, 47 ff.
    heat of, 25 ff., 36, 50–52, 55
    magnetism of, 38, 45 ff., 52, 109 ff.
    mantle of, 43, 51, 52, 53, 54, 57
    origin of, 25 ff.
    revolution of, 17, 68–69
    rotation of, 16, 17, 22, 40, 48, 64, 70 ff., 75, 87, 88, 89 ff.
    shape of, 9, 16, 21 ff.
    structure of, 26, 48 ff.
    weight of, 18

Index

Earthquakes, 48 ff., 54, 56, 60
Electricity, 52, 111. *See also* Current, electric
Electrojet, 109, 113
Electrons, 48, 51–52, 104, 108, 109, 111, 112, 113, 115
Elements, 25, 30–32
  radioactive, 25, 28–29, 117
Equator, 22, 41, 86, 87, 88, 89, 109
Eratosthenes, 16
Exosphere, 101, 106 ff.
Explorer, 101, 112

F region, 101, 105
Ferrel, William, 90
Ferrel's law, 89 ff.
Fisher, Osmond, 23, 42
Flood, the, 33, 39

Galilei, Galileo, 18
Gamma-rays, 108, 117
Gauss, Karl, 9
Geophysical Observatory, Orbiting, see OGO
Geophysical Year, International, 11 ff., 15, 61, 62, 99
Geophysics, definition of, 7
  applied, 13 ff.
Georghios, Stotto, 75
Geosyncline, 34, 44
Geothermal gradient, 27
'GHOST', 95
Giekie, Archibald, 28
Goddard, Robert H., 11
Gondwanaland, 38, 39, 43, 46, 48
Granite, 36, 55, 58
Gravitation, 36, 38, 40, 42, 67, 109
Gravity, 14, 17, 22, 25, 48
Greely, Adolphus, 10
Green, Lowthian, 22, 23
Greenland, 10, 15, 41
Griggs, David, 44
Gulf Stream, 65, 120
Gutenberg, B., 52

Hadley, George, 89, 90, 91
Halley, Edmund, 92
Halm, J. K. E., 47
Heat equator, 86, 88
Heat flow, 27, 36, 44, 74
Heaviside, Oliver, 104
Heim, Albert, 34
Helium, 30, 31, 77, 78, 85, 107, 108
Heraclides, 16
Hermite, C., 94
Hipparchus, 17
Hopkins, William, 42
Horse Latitudes, 87, 88
Houet, G. S., 83
Humboldt, Alexander von, 9
Huygens, Christiaan, 22
Hydrogen, 25, 30, 31, 85, 105, 107, 108, 117, 118
Hydrosphere, 18, 57 ff.
Hydroxyl ions, 105, 118
  layer, 101, 106

Idrac, P., 94
IGY, *see* Geophysical Year
Ionosphere, 100, 101, 104–105, 109, 113
IQSY, 12
Iron, 31, 37, 38, 50, 52, 53, 61, 118
Islands, arcs (festoons) of, 41, 59
  continental, 59
  oceanic, 59
Isostasy, 37, 119

Jeffreys, Harold, 22, 40, 43
Jet stream, 89, 100
Joly, John, 36

Keller, H., 80
Kelvin, Lord, *see* Thomson, William
Kennelly, A. E., 104
Kennelly-Heaviside layer, 104
Kepler, Johannes, 18
Kipfer, Paul, 96

Lamarck, Jean Baptiste, 8
Laplace, Pierre Simon, 8
Latham, Gary, 75
Laurasia, 43, 48
Lavoisier, Antoine, 8
Lead, 28–29, 31, 32
Leibniz, Gottfried von, 26
Light rays, 100, 103
Lindbergh, J., 78
Link, E., 77, 78
Lippershey, Johannes, 18
Lithosphere, 18, 33 ff.
Love, A. E. H., 49

McClure, Clifton, 97
Magma chamber, 44, 55, 56
Magnesium, 31, 37, 53, 63, 118
Magnetic phenomena, 11, 12, 46, 52, 109, 112 ff.
  rocks, 45–47
  survey, 9
Magnetosphere, 109 ff.
Manganese, 31, 61, 62
Manhigh, 97
Marconi, G., 104
Marianas Trench, 27, 57, 81, 83
Maupertuis, P. L. M. de, 22
Maury, Matthew, 8
Maxwell, A. E., 74
Mesopause, 100
Mesosphere, 99
Meteor, 25, 102, 105
  graveyard, 101, 108
Mine, deepest, 26
'Mohole', 53 ff.
Mohorovičić, A., 52
Monsoon, 93
Moon, 16, 21, 23–24, 40, 41, 67, 68–70, 72, 109, 118
Mountains, fold, 34
  fore-edge, 44
  formation of, 33 ff., 40 ff.
  height of, 19, 26–27
  submarine, 59, 61

Newton, Isaac, 18, 21–22
Nickel, 31, 38, 50, 61, 102
Nife, 38
Night air glow, 101, 105
Nitrogen, 31, 76, 77, 78, 85, 105, 106, 108, 115, 117
North Atlantic Drift, 65
Nutation, 42 *n.*

Oceans, 57 ff., 84
  area of, 57
  bottom of, 58 ff., 74 ff.
  depth of, 18, 19, 27, 57–58, 73–74
OGO, 101, 111
Oldham, R. D., 49, 52
Orogenesis, 34
Oxygen, 30, 31, 77, 78, 85, 105, 106, 108, 115, 117–118
Ozone, 85, 117
  layer, 100, 101, 103, 106

Pacific, 23–24, 28, 55, 59, 62, 66, 70
Pangaea, 41, 48
Pascal, Blaise, 94
Pepys, Samuel, 102
Pianstanida, N., 96
Piccard, Auguste, 83, 96
  Jacques, 83
Placet, P., 39
Polar calms, 87
  compression, 22, 24, 25
  electrojet, 113
  front, 87, 89
  lights, 115
  year, 9 ff.
Pole, North, 10, 25, 41, 42, 90
  secular, 42 *n.*
  South, 12, 25, 40, 41, 42
Poles, 22, 40
  'flight from the', 41
  magnetic, 45 ff., 110, 112–113
  'wandering of the', 42

Index

Pratt, J. H., 36
Precession of the equinoxes, 42 n.
Pressure, atmospheric, 63, 85, 86, 94, 96, 100, 108
  in interior of earth, 48, 50–52, 55–56
  submarine, 63–64, 76 ff.
Prospecting, geophysical, 13 ff., 80
Protons, 111, 112, 113, 115
Ptolemy of Alexandria, 17
Pythagoreans, 16

Radar, 14, 106, 108
Radiation belts, 101, 110, 112, 113
Radio-sonde, 11, 94–95
Radio transmission, 11, 101, 104, 105
Radioactivity, 25, 28–29, 36, 47, 117
Rain belt, equatorial, 86–87
Rayleigh, Lord, 49. *See also* Strutt, Robert John
Revolution, 35, 43
Ridge, 44, 59, 60
Rift, 44, 45, 61
Rigaud, L. de, 77
Rise, *see* Ridge
Roaring Forties, 87, 89
Rocket, 11, 98–99, 101, 102
'Rockoon', 99
Rocks, crustal, 29, 59, 61. *See also* Earth, crust of
  igneous, 28, 36 ff.
  magnetic, 13, 45–47
  molten, 36, 42, 45, 50, 51, 54, 55, 56,
  olivine, 37, 53
  organic, 119
  plasticity of, 36, 39–40, 42 ff.
  primeval, 118
  sedimentary, 28, 34, 44

St. Gotthard Tunnel, 40
Salinity, 65–66, 74

Sargasso Sea, 65
Satellite, artificial, 19, 101, 105, 107, 108, 111, 112, 119
  geodetic, 24
Schuster, A., 109
Sea-floor Geophysical Recording Station, 75
Sea-floor spreading, 47, 60
Sealab, 78, 79
Seamount, 59
Sea-water, colour of, 62–63
  composition of, 63
  temperature of, 64
Shelf sea, 58, 59
Sial, 37, 39, 47, 52, 55, 57, 58
Siebe, Augustus, 76
Silica, 36 ff.
Sima, 37, 39, 52, 55, 57, 58
Skylark, 98
Small, P., 80
Snider, Antonio, 39
Sodium, 31, 37, 63
  layer, 101, 105, 106
  vapour, 106
Solar flare, 113
  wind, 110, 111–112
Sonar, 74
Sound, 100, 102–103
  speed in water, 74
Sounding, deep-sea, 73–75
  echo, *see* Sonar
  balloon, *see under* Balloon
'SPID', 78
Sputnik, 101
Sténuit, R., 78
Stewart, Balfour, 109
Stratosphere, 19, 96–97, 100, 103
Strutt, Robert John, 29
Suess, Eduard, 34, 37, 38, 39, 47
Sun, 12, 16, 17, 23–24, 25, 30, 67, 68, 86, 105, 109, 111, 113, 118
Sunspot, 12, 26

Taylor, F. B., 40

Tektite I, 79
Temperature, of air, 86–89, 94, 96, 99, 100–101, 102–103, 105, 107, 108
   of earth, *see* Earth, heat of
   of sea, 64, 66
Tethys, 34, 38
Tetrahedral hypothesis, 22–23
Thermosphere, 99, 100, 101, 102, 105
Thomson, William, 21, 28
Thorium, 15, 29, 31
Tides, 67 ff.
   atmospheric, 109
   highest, 73
Toit, A. L. du, 43
Torricelli, Evangelista, 94
Trade Winds, 65, 87, 88, 89, 93
Trenches, 58, 59, 60
Trieste, 82, 83
Tropopause, 96, 100
Troposphere, 85 ff., 96, 100, 103
Tsunami, 70 *n.*, 75
Twilight flash, 105

Ultra-violet rays, 101, 103, 117
Uranium, 15, 28–29, 31

Vegetation, 20, 118–119
Vinci, Leonardo da, 33
Vityaz, 58
Volcano, 26, 45, 54, 55, 56, 88

Walsh, D., 83
Water vapour, 85, 86, 96, 117, 118
Waves, earthquake, 49–50, 51
   oscillating, *see* standing
   progressive, 71, 72
   radio, 101, 104–105
   sound, 102–103
   standing, 70, 71, 72, 73
   tidal, 70 ff.
Weather, 8, 85, 89, 100, 103, 119
Weber, Wilhelm, 9
Wegener, Alfred, 40 ff., 47
Westerly Variables, 64, 65, 87, 88 89
Weyprecht, Karl, 9
Wilkins, Bishop, 77
Willm, P. H., 83
Wilson, Alexander, 94
Wind system, planetary, 87 ff.

X-rays, 104, 113, 117

Zodiacal light, 112